PEARSON

Math Makes Sense 2

Author Team

Sandra Ball
Maggie Martin Connell
Lori Jane Hantelmann
Sharon Jeroski
Peggy Morrow
Carole Saundry
Mignonne Wood

With Contributions from

Maureen Dockendorf
Linden Gray
Susan Green
Michelle Jackson
Jill Norman
Heather Spencer

PEARSON

Publisher
Mike Czukar

Research and Communications Manager
Barbara Vogt

Publishing Team
Claire Burnett
Lesley Haynes
Cristina Getson
Keltie Thomas
Ellen Davidson
Jane Schell
Lynda Cowan
Jon Maxfield
Margaret McClintock
Lynne Gulliver
Cheri Westra
Judy Wilson

Design and Art Direction
Word & Image Design Studio Inc.
Carolyn Sebestyen

PEARSON

ISBN-13: 978-0-321-46929-8
ISBN-10: 0-321-46929-1

Printed and bound in the United States

8 9 10 11 12 15 14 13 12 11

Acknowledgments
The publisher wishes to thank the following sources for photographs, illustrations, and other materials used in this text. Care has been taken to determine and locate ownership of copyright material in this book. We will gladly receive information enabling us to rectify any errors or omissions in credits.

Cover
Cover illustration by Marisol Sarrazin

Illustrations
Kasia Charko, 105; Jeff Dixon, 15; Virginie Faucher, 175, 182; Marie-Claude Favreau, 1-9, 89-100, 155-162, 205-212, 216; Joanne Fitzgerald, 102, 103 (right); Leanne Franson, 125; Linda Hendry, 30 (bottom), 33-35, 37, 38, 43, 44-47, 50, 51, 55, 57, 171, 195, 199, 200 (top); Tina Holdcroft, 10-12, 17, 18, 20, 23-26, 28, 48, 49, 106, 107, 109, 111-114, 116-119, 123, 126, 127, 150, 151, 153, 154, 178, 185, 187; Vesna Krstanovic, 29, 30 (top), 31, 32, 36, 39-42, 53, 54, 58, 62-75, 77-81, 83-87, 115, 122, 124, 190-193, 197, 198, 200 (bottom)-204; Andre Labrie, 14, 16, 19, 21, 22, 27; Bernadette Lau, 146; Derek Matthews, 13; Paul McCusker, 108, 110, 141; Allan Moon, 176, 177, 180, 181; Scot Ritchie, 59-61, 76, 82, 103 (left), 120, 121, 128-131, 133, 134, 136-140, 142, 143, 145, 147-149, 152, 167-170, 218-220; Bill Slavin, 172, 186, 188; Laurie Stein, 135, 144; Pat Stephens, 132

Photography
Ray Boudreau, 101 (left), 102 (left), 104, 167 (right), 217 (right); Ian Crysler, 101 (right), 167 (left), 219 (right); Shutterstock, 189, with the exception of Jupiterimages Unlimited: bottom left 7 items windpipes to accordion, top left corner, and top to right of drums

Contents

Consultants, Advisers, and Reviewers

Series Consultants
Trevor Brown
Maggie Martin Connell
Craig Featherstone
John A. Van de Walle
Mignonne Wood

Assessment Consultant
Sharon Jeroski

Cultural Consultant
Ken Ealey

Editorial Consultants
Lalie Harcourt
Ricki Wortzman

Advisers and Reviewers

Pearson Education thanks its advisers and reviewers, who helped shape the vision for *Pearson Mathematics Makes Sense* through discussions and reviews of prototype materials and manuscript.

Alberta

Lorna Addison
Calgary Board of Education

Joanne Adomeit
Calgary Board of Education

Bob Berglind
Formerly Calgary Board of Education

Jacquie Bouck
Lloydminster Public School Division 99

Auriana Burns
Edmonton Public School Board

Daryl Chichak
Edmonton Catholic School District

Lorelee Clack
Calgary Board of Education

Lissa D'Amour
Medicine Hat School District 76

Brenda Foster
Calgary R.C.S.S.D. 1

Nina Fotty
Edmonton Public School Board

Florence Glanfield
University of Alberta

Ellen Guderyan
Calgary R.C.S.S.D 1

Candace Krush
Calgary Board of Education

Jodi Mackie
Edmonton Public School Board

Korilee Marks
Golden Hills School Division

Kathy Monar
Calgary Board of Education

Cathy Plastow
Edmonton Public School Board

Jeffrey Tang
Calgary R.C.S.S.D. 1

Bonnie Terlesky
Edmonton Public School Board

British Columbia

Sandra Ball
Surrey School District 36

Lorraine Baron
Central Okanagan School District 23

Donna Beaumont
Burnaby School District 41

E. Jane Cowan
Formerly West Vancouver

Marc Garneau
Surrey School District 36

Selina Millar
Surrey School District 36

Kathleen Sarton
Greater Victoria School District 61

Chris Van Bergeyk
Central Okanagan School District 23

Denise Vuignier
Burnaby School District 41

Mignonne Wood
Formerly Burnaby School District 41

Manitoba

Rosanne Ashley
Winnipeg School Division

Angela M. Bubnowicz
Seven Oaks School Division

Cheryl Horsfall
Pembina Trails School Division

Ralph Mason
University of Manitoba

Christine Ottawa
Mathematics Consultant, Winnipeg

Gretha Pallen
Formerly Manitoba Education

Sandy Smith
Kelsey School Division

Gay Sul
Frontier School Division

Wendy Weight
Pembina Trails School Division

Saskatchewan

Susan Beaudin
File Hills Qu'Appelle Tribal Council

Edward Doolittle
First Nations University, U of Regina

Lori Jane Hantelmann
Regina School Division 4

Rayleen Eberl
Prairie South School Division

Angie Harding
Regina R.C.S.S.D. 81

Wilma Mantei
S.E. Cornerstone School Division

Cheryl Shields
Spirit School Division

Cindy Toniello
Regina School Division 4

School Begins

“It’s time for school,” Cam’s grandma said.
“Open your eyes. Get out of bed.
I’m coming to school today with you,
because you’re starting somewhere new.”

"I'm scared," Cam said, then chewed his toast.
He hated changing schools the most.
"I won't have friends."
Then Grandma sighed.
"You'll make *new* friends, Cam," she replied.

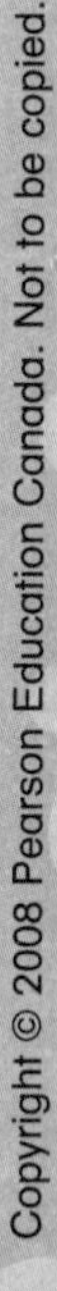

The teacher greeted them at the door,
and welcomed Cam to Classroom 4.
The chairs and tables were tightly fit.
Cam looked around. "Where should I sit?"

The teacher pointed. "There's a chair.
You can sit with those children there.
Your grandma can sit at that table, too,
if she'd like to stay awhile with you."

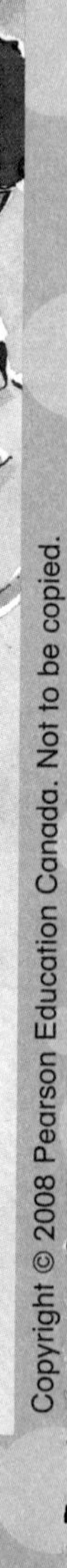

The teacher asked Grandma, “Can you help today?
We’re learning math games we can play.
There are materials to share and rules to learn,
like when to move and take a turn.”

The class listened to the teacher explain.
She repeated some of the rules again.
She said, “Do you have questions? Raise your hand.”
But the class said, “No, we understand.”

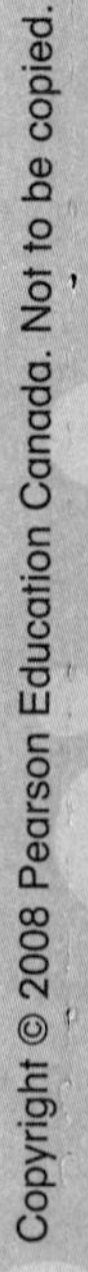

At recess, Grandma had to go.
She said, “This is the best Grade 2 I know!”
She waved at all her new young friends,
who called, “Come back, and help again!”

About the Story

The story was read in class to prepare for a Mathematics Investigation activity. Children played a variety of mathematical games and created pattern and number stories. The Investigation provided opportunities for the teacher to learn about children's mathematical understanding and skills as they begin a new school year.

Talk about It Together

- How did Cam feel about his first day of school?
- What do you think Cam's math class will be like? Why?
- What did Cam's teacher do to help the children work together?
- How is Cam's classroom the same as your classroom? How is it different?

At the Library

Ask your local librarian about books with math-related themes for Grade 2 readers.

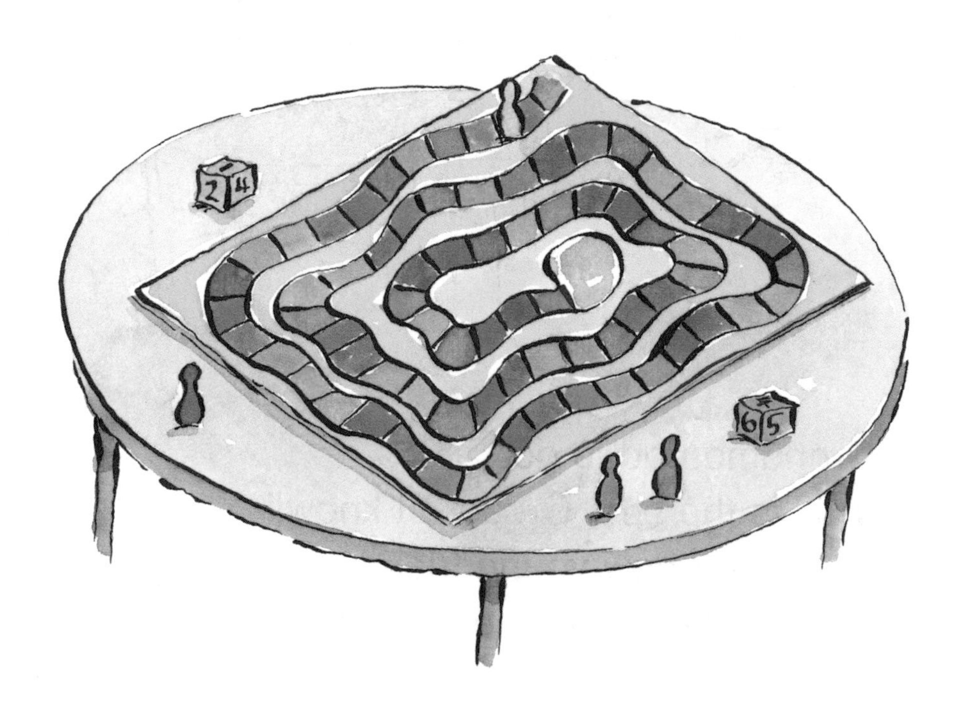

Grandma Helps

Make a number story about the picture.
Use pictures, numbers, or words to tell your story.

Show the Same Pattern in Different Ways

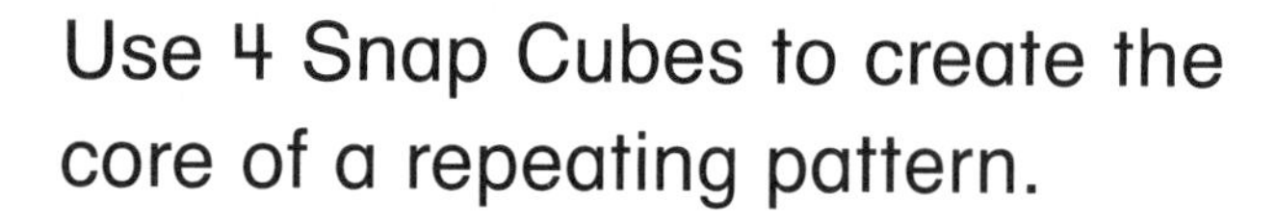

Use 4 Snap Cubes to create the core of a repeating pattern.

Colour the cubes to show 3 repeats of your pattern.

Show your pattern using a letter code.

___ ___ ___ ___ ___ ___ ___ ___ ___ ___ ___ ___

Mark the core in each way you showed your pattern.

How are the ways of showing your pattern the same?

How are the ways of showing your pattern different?

Building Challenge!

Use a spinner.

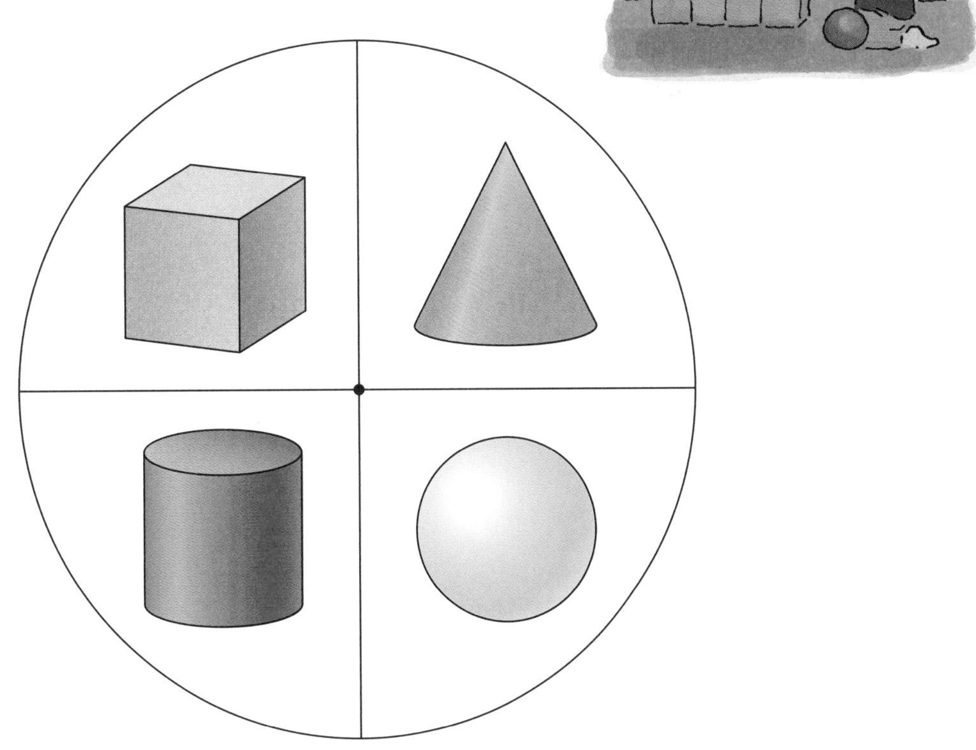

Spin 10 times.
Each time you spin, take the object you landed on.
Build the tallest structure you can.
Tell how many of each you have.

3-D Object	How Many I Have	How Many I Used
cone		
cylinder		
sphere		
cube		

A Special Game

Make your own math game.
You can use spinners, number cubes, counters, objects, or anything you like.

Tell how to play your game.

What do you need to play your game?

Write the rules to your game.

Teach other children how to play your game.

Patterning

Focus | Children discuss the train car patterns, and talk about how the train cars are the same and different.

Name: ______________________ Date: ______________________

Dear Family,

Your child is starting a unit in mathematics on patterning.

The Learning Goals for this unit are to

- Describe, extend, compare, and draw repeating patterns and increasing patterns.
- Create new patterns.
- Talk about a pattern rule.
- Use 2 attributes to make a pattern.

You can help your child achieve these goals by doing the Home Connection activities suggested at the bottom of selected pages.

Name: ______________________ Date: ______________________

The Same or Different?

Circle the train car in each row that is the same as the first car.

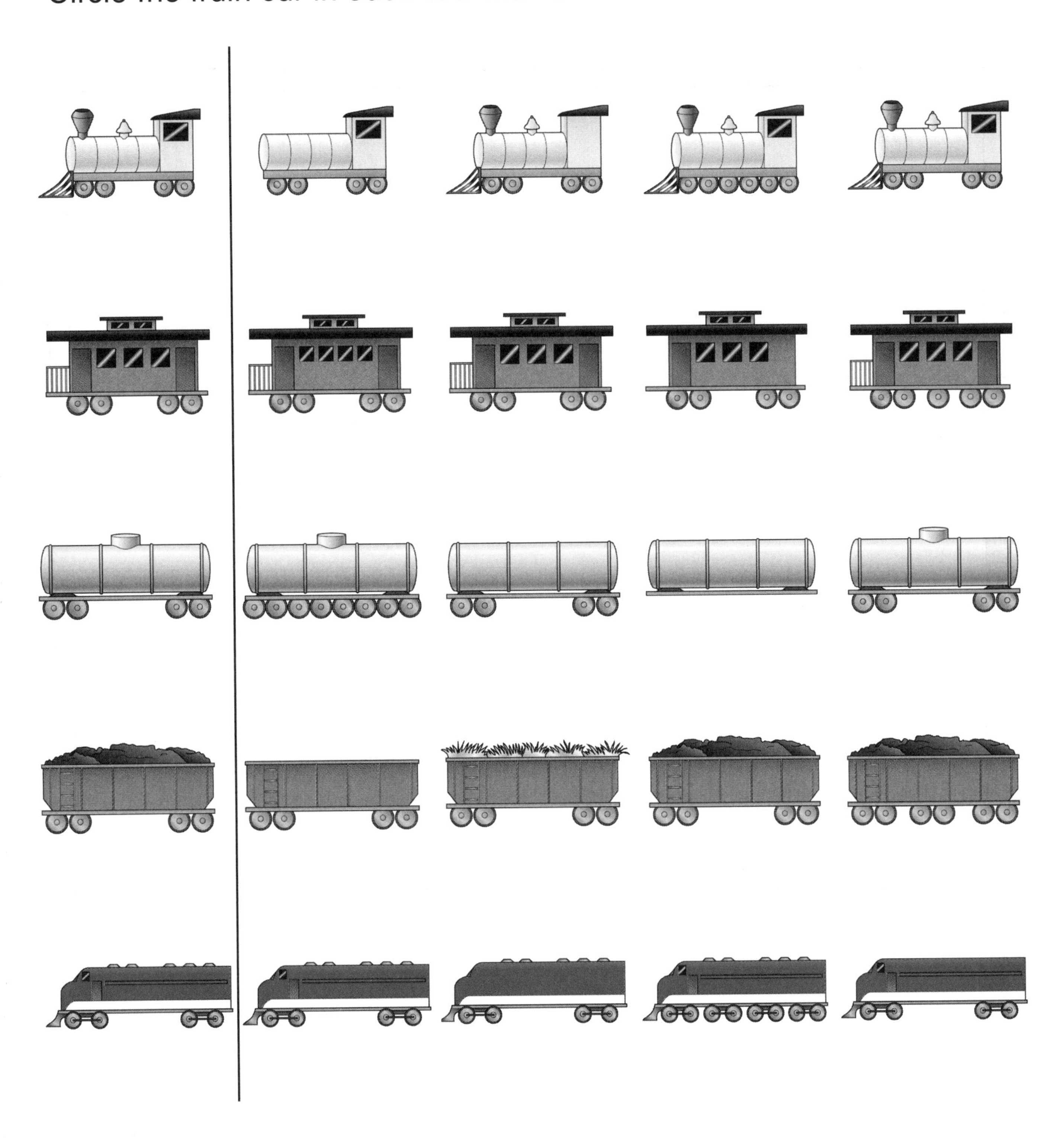

Focus | Children find the train car that is the same as the first car in each row.

Name: ______________________ Date: ______________________

Describe, Extend, Compare

Draw your pattern.
Extend it.

What is the pattern core?

____

HOME CONNECTION 

With your child, look around your home for patterns. Ask your child to draw a picture to represent each pattern.

Focus | Children record the patterns they have made, and extend them.

Name: ______________________ Date: ______________________

Extend the Pattern

Use the cutouts from Line Master 2
to extend the patterns.
Colour the cutouts to match the patterns.

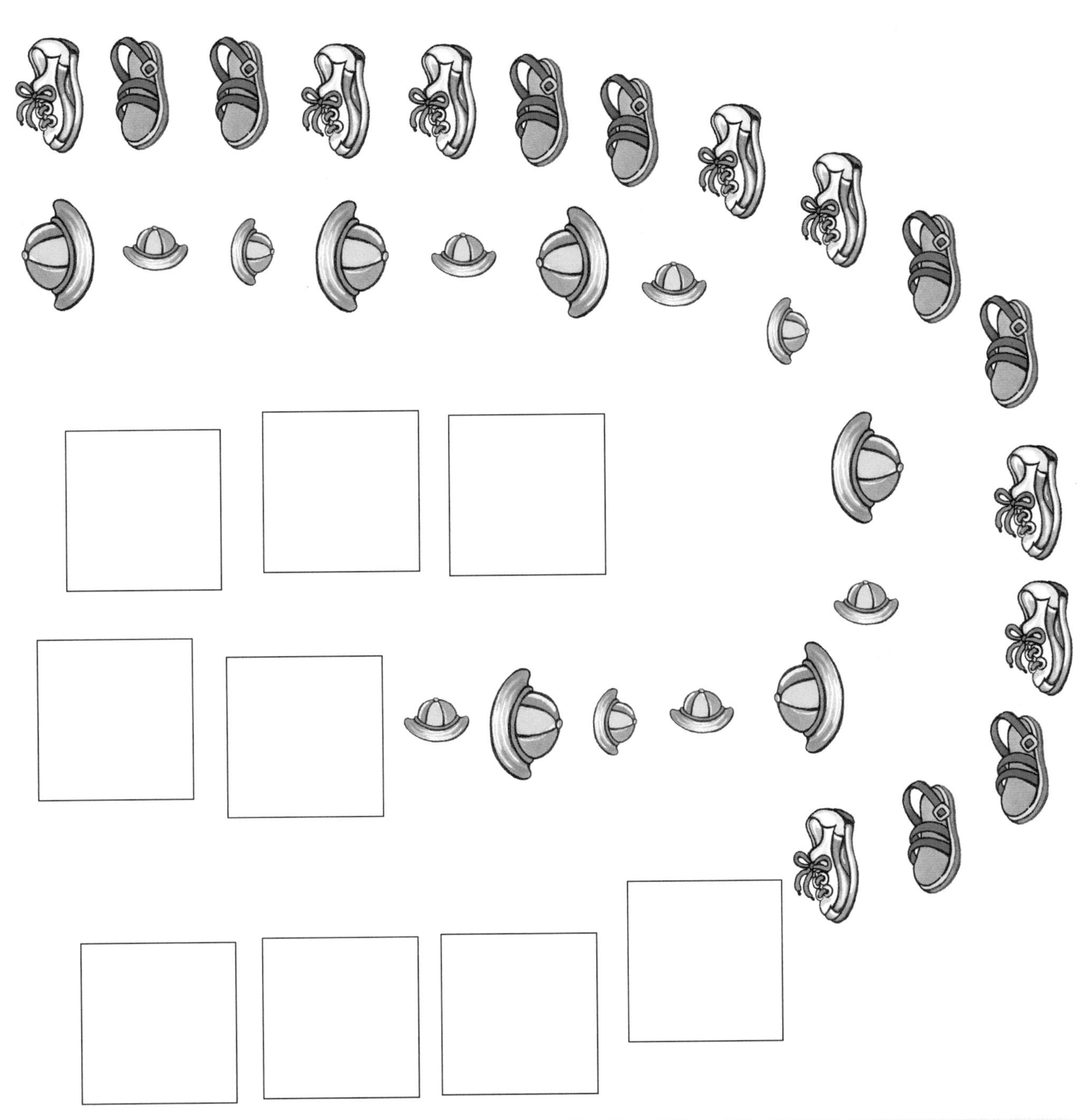

Focus | Children extend the given patterns.

Name: ______________________ Date: ______________________

What Is Different?

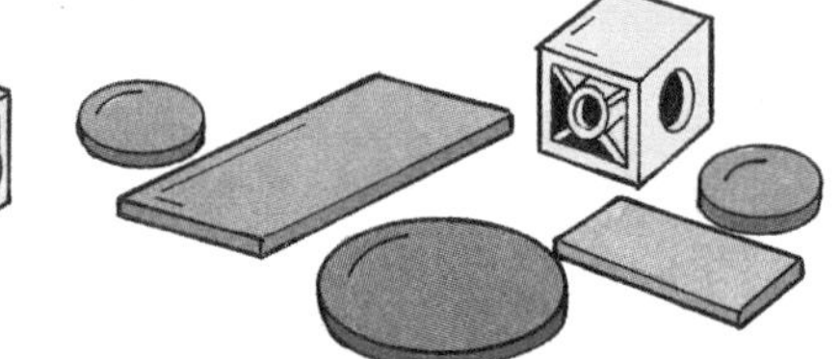

Circle the core in each pattern.
What are the 2 attributes in each pattern?

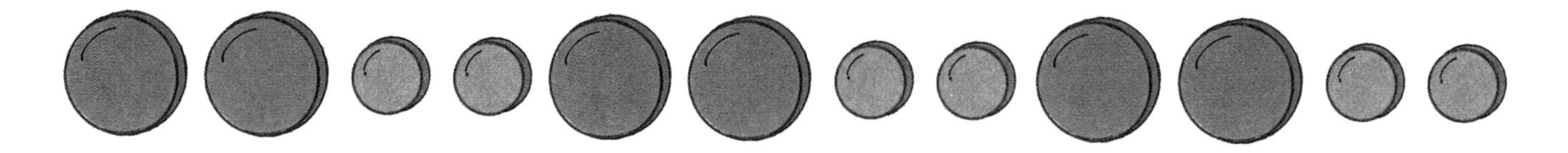

______________________ ______________________

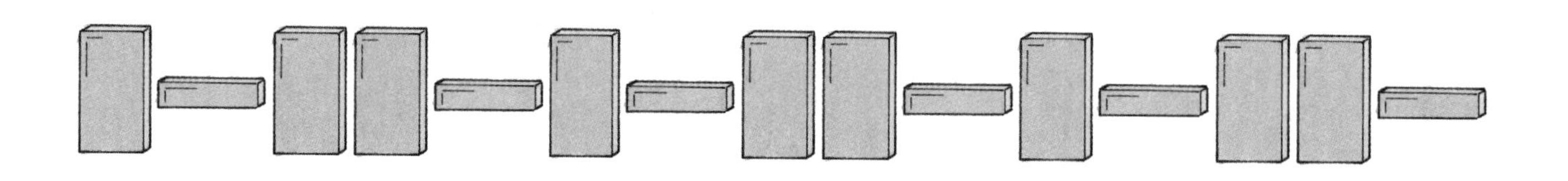

______________________ ______________________

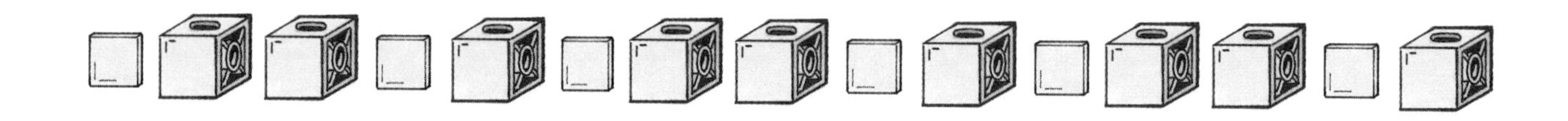

______________________ ______________________

Make a pattern with 2 attributes. Draw it here.

FOCUS | Children identify 2 attributes in various patterns. Then, they make their own pattern.

Name: ______________________ Date: ______________________

Making Patterns

Make a pattern. Use △ and □.

Look at a friend's pattern. How are the patterns the same?

____

How are the patterns different?

____

How did you make your pattern?

____

____

Focus | Children make and compare patterns with 2 attributes. There are many correct answers.

HOME CONNECTION 

Have your child use spoons (large and small) to make a pattern 2 different ways. Ask: "How are the patterns the same? How are they different?"

Name: ______________________ Date: ______________________

What Is Missing?

Draw and colour each missing shape.

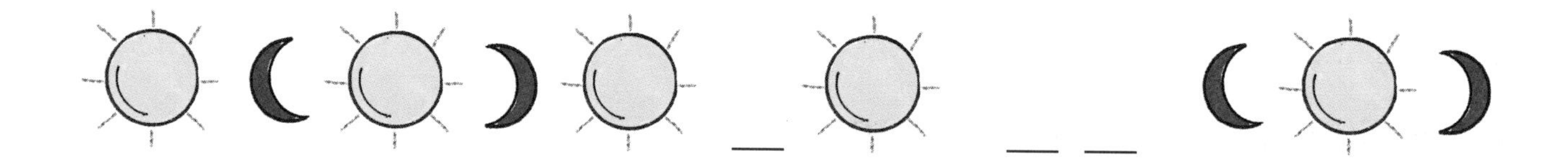

HOME CONNECTION 

Make a pattern (plate, spoon, mug, spoon) and repeat the core 3 times. Remove some items and ask your child, "What's missing?" Then ask, "Where does the next spoon go?"

FOCUS | Children fill in what is missing in each pattern.

Name: ______________________ Date: ______________________

Use All the Beads

There are 6 red, 3 green,
and 6 blue beads.

Make a pattern with the beads.
Use all the beads.

Show your thinking in pictures or words.

Focus | Children use all the beads to make a pattern. There are many correct answers.

Name: ______________________ Date: ______________________

Bead Pattern

There are 3 brown, 6 yellow, and 6 purple beads.

Make a pattern.
Use all the beads.

Show your thinking in pictures or words.

Focus | Children use all the beads to make a pattern. There are many correct answers.

HOME CONNECTION 

Ask your child: "How did you make your pattern?"

Name: ______________________ Date: ______________________

What Is the Rule?

What comes next? Extend the pattern.

What is the pattern rule?

____

What comes next? Extend the pattern.

What is the pattern rule?

____

HOME CONNECTION 

Look around your home with your child for patterns. Ask your child, "Is this an increasing pattern? Why (or why not)?"

FOCUS | Children copy, extend, and describe increasing patterns.

Name: ______________________ Date: ______________________

Increasing Patterns

Extend the pattern. Show 1 more element.

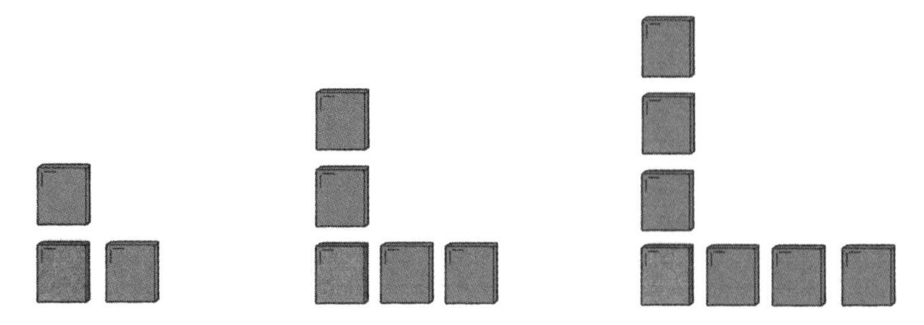

Show the same pattern using different objects, letters, or numbers.

Make your own increasing pattern.
Use the same starting element, and make a different pattern.

FOCUS | Children extend and reproduce a given increasing pattern using another mode, then create their own increasing patterns.

Name: ______________________ Date: ______________________

How Many Drumsticks?

How many drumsticks would 10 drummers have?

Show your thinking using pictures, numbers, or words.

HOME CONNECTION 

Ask your child, "How many socks would we need for everyone in our family?" Then collect a pair of socks for everyone, and count the socks.

FOCUS | Children extend an increasing pattern to show how many drumsticks 10 drummers would have.

Name: ______________________ Date: ______________________

Counting Wheels

How many wheels do 6 tricycles have? _____

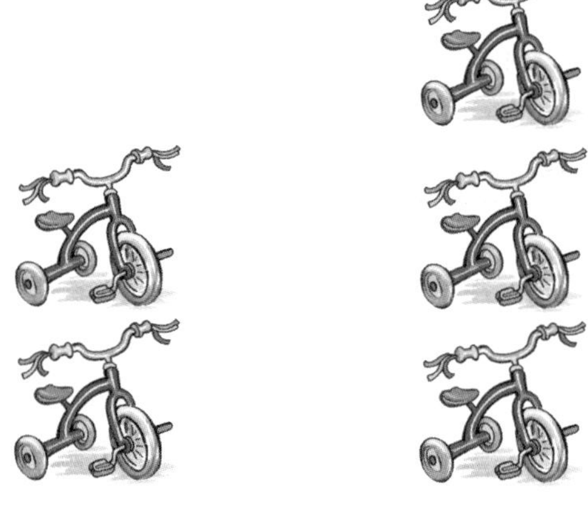

Tell how you solved the problem.
Use pictures, numbers, or words.

Focus | Children extend an increasing pattern to help them determine the number of wheels on 6 tricycles. They record their pattern and show how they solved the problem.

Name: ______________________ Date: ______________________

My Increasing Pattern

Draw an increasing pattern.
Describe it to a partner.

HOME CONNECTION 

With your child, make an increasing pattern using crayons, markers, stickers, pencils, or other favourite objects. Change the order of objects and make another pattern.

Focus | Children draw and describe their increasing pattern.

Name: ______________________ Date: ______________________

My Journal

Show what you learned about using 2 attributes to make a repeating pattern.

Show what you learned about making an increasing pattern.

Focus | Children reflect on what they have learned about repeating and increasing patterns.

Numbers to 100

Focus | Children talk about the picture and identify numbers of objects.

Name: ______________________ Date: ______________________

Dear Family,

This unit will focus on deepening your child's understanding of number relationships, counting, and place value.

The Learning Goals for this unit are to

- Read numbers to 100 in symbols and in words.
- Build numbers to 100 with concrete materials.
- Estimate the number of objects in groups to 100.
- Count forward and backward using different tools, such as objects, a number line, a 100-chart, and coins.
- Show numbers to 100 as tens and ones using a variety of materials.
- Compare and order numbers to 100.

You can help your child achieve these goals by doing the Home Connection activities suggested at the bottom of selected pages.

Name: ______________________ Date: ______________________

How Many?

<table>
<tr>
<td>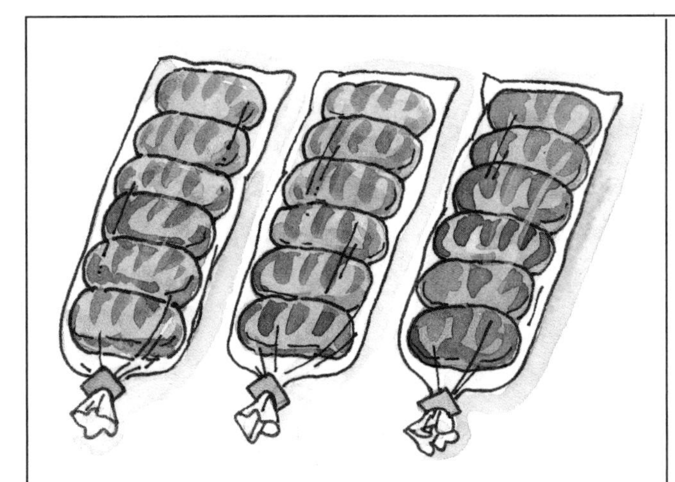
How many rolls? ________</td>
<td>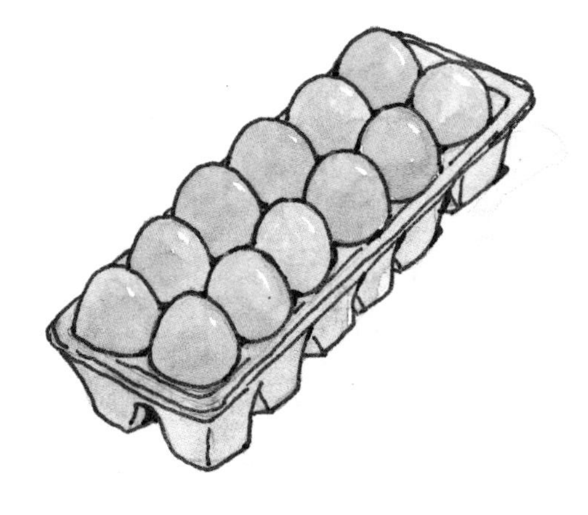
How many eggs? ________</td>
</tr>
<tr>
<td>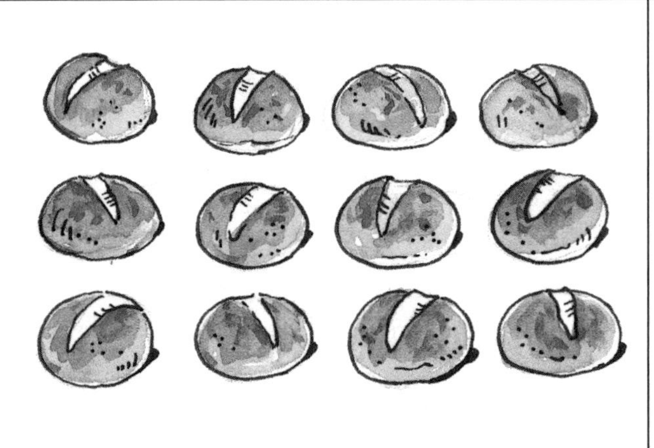
How many buns? ________</td>
<td>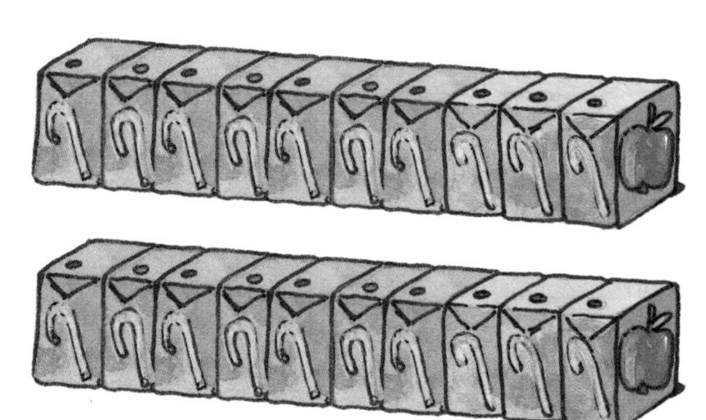
How many boxes? ________</td>
</tr>
</table>

Which groups of items show the same number?

____

How do you know?

____

Focus | Children count and record the number in each group and determine which groups show the same number.

Name: ______________________ Date: ______________________

What Is Missing?

Write the missing numbers on the number lines.

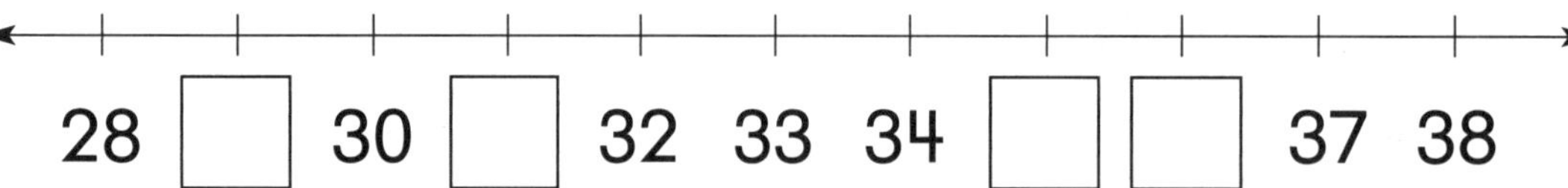

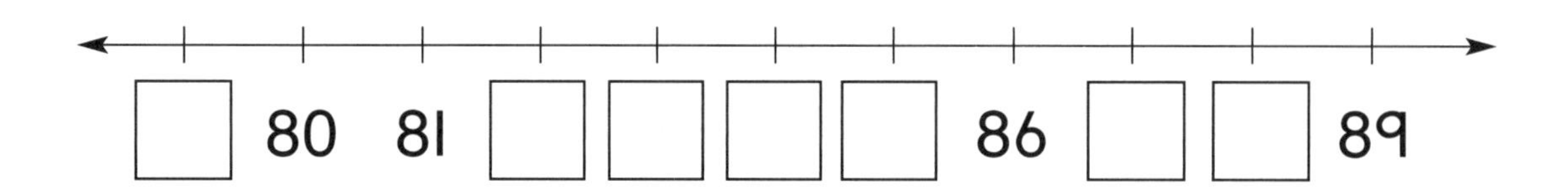

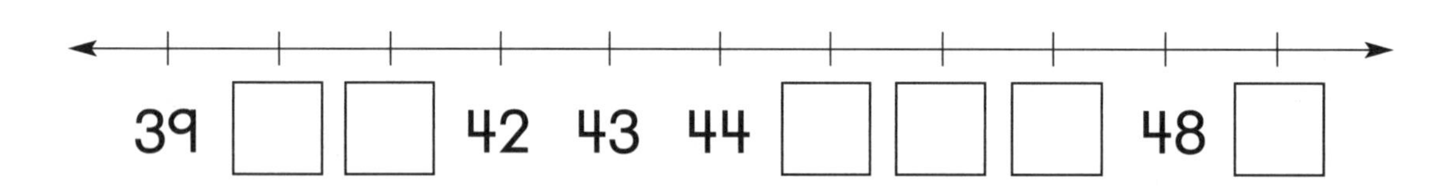

Count forward by 5s from 0 to 30. Write the numbers.

____

Count backward by 10s from 90 to 30. Write the numbers.

____

Count forward by 2s from 82 to 94. Write the numbers.

____

Count backward by 5s from 60 to 35. Write the numbers.

____

Focus | Children complete number lines and count forward and backward.

HOME CONNECTION 

Help your child count forward and backward by 2s, 5s, and 10s.

Name: ______________________ Date: ______________________

Will You Say 35?

Show your thinking in pictures, numbers, or words.

Start at 10 and count forward by 2s.
Will you say 35?

Start at 65 and count backward by 5s. Will you say 35?

Start at 100 and count backward by 10s. Will you say 35?

Focus | Children use counting patterns to solve problems.

Name: ______________________ Date: ______________________

Counting by 2s, 5s, and 10s

What is the counting pattern?

55, 60, 65, 70, 75 ______________________

90, 80, 70, 60, 50 ______________________

7, 17, 27, 37, 47 ______________________

94, 92, 90, 88, 86 ______________________

Write the missing numbers.

100, 95, 90, ______, ______, ______, 70, ______, ______

74, 76, 78, ______, ______, 84, ______, ______, ______

3, 13, 23, ______, 43, ______, ______, ______, 83

1, 3, 5, ______, ______, ______, 13, ______, ______

Here is a counting by 5s pattern with some errors.

5, 10, 15, 25, 25, 30, 35, 30, 45, 50, 60

Write the correct counting pattern.

____

____

Focus | Children identify, extend, and correct errors in counting patterns.

Name: ______________________ Date: ______________________

Will You Say 49?

Show your thinking in pictures, numbers, or words.

Start at 30 and count forward by 5s. Will you say 49?

Start at 1 and count forward by 2s. Will you say 49?

Start at 9 and count forward by 10s. Will you say 49?

Focus | Children use counting patterns to solve problems.

Name: ______________________ Date: ______________________

Odd and Even Numbers

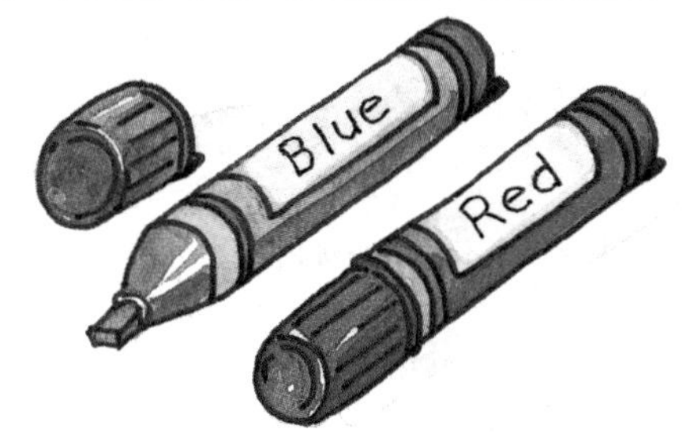

Colour the even numbers from 50 to 68 red.
Colour the odd numbers from 19 to 37 blue.

1	2	3	4	5	6	7	8	9	10
11	12	13	14	15	16	17	18	19	20
21	22	23	24	25	26	27	28	29	30
31	32	33	34	35	36	37	38	39	40
41	42	43	44	45	46	47	48	49	50
51	52	53	54	55	56	57	58	59	60
61	62	63	64	65	66	67	68	69	70
71	72	73	74	75	76	77	78	79	80
81	82	83	84	85	86	87	88	89	90
91	92	93	94	95	96	97	98	99	100

What patterns do you see?

____

____

____

Focus | Children colour a 100-chart to show odd and even numbers.

HOME CONNECTION 

Have your child use the chart on this page to describe number patterns on a 100-chart.

Name: ______________________ Date: ______________________

Counting with Coins

Put the coins in each wallet.
Write the amount of money in each wallet.

_______¢ in all

_______¢ in all

_______¢ in all

_______¢ in all

_______¢ in all

_______¢ in all

HOME CONNECTION 

Place 6 pennies in a row. Ask your child: "How many pennies are there?" Add a dime to the row and ask: "How much money is there altogether?" Continue using dimes to count until you reach 96¢.

Focus | Children use coins to find money amounts up to 100¢.

Name: ______________________ Date: ______________________

Ordinal Numbers

Jolene's locker is the 9th locker in this hallway.

What letter is on Jolene's locker? _______

What letter is on the 3rd locker? _______

What letter is on the 7th locker? _______

What letter is on the 10th locker? _______

Which ordinal number is the runner in the striped shirt? ______

Which ordinal number is the runner in the striped shirt now? ______

HOME CONNECTION 

Line up 10 things in a row, such as spoons, forks, cups, or plates. Ask your child: "What is third? Fifth? Ninth?"

Focus | Children describe order or relative position using ordinal numbers (up to tenth).

Name: ______________________ Date: ______________________

Estimating and Counting

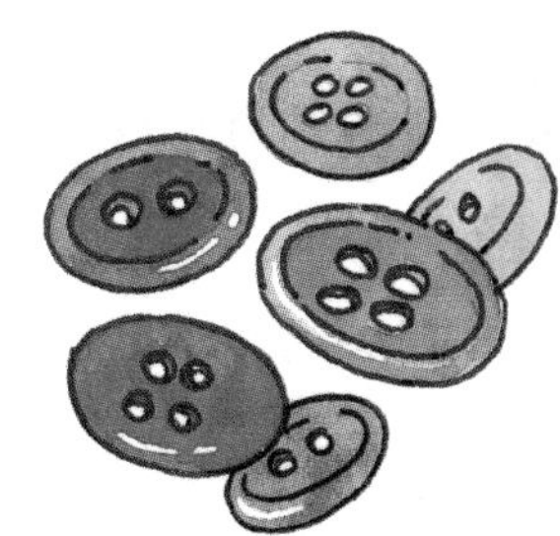

Spill the objects. Estimate the number.__________

Tell how you estimated.

Use pictures, numbers, or words.

Count the objects.

Tell how you counted.

Use pictures, numbers, or words.

Focus | Children estimate the number in a collection of up to 50 objects and then count the objects. They describe how they estimated and how they counted using pictures, numbers, or words.

Name: ______________________ Date: ______________________

How Many Buttons?

Estimate the number of buttons. _______

Tell how you estimated. Use pictures, numbers, or words.

How many buttons are there? _______

Tell how you counted. Use pictures, numbers, or words.

HOME CONNECTION 

Gather a collection of about 40 small objects for your child to count, such as toys or pennies. Ask your child to count the collection by grouping the objects in different ways.

Focus | Children estimate, then count a collection using a strategy of their choice.

Name: ______________________ Date: ______________________

How Many Do You Have?

Take a handful of small objects with both hands.
Circle an estimate for the number of your objects.

40 80

Tell how you chose the estimate.
Use pictures, numbers, or words.

Place the objects on ten frames.
Colour the pictures of ten frames to match your ten frames.

________ groups of 10 and ________ left over ________ in all

Focus | Children take a handful of objects and choose an estimate. Then, they place the objects on the ten frames and count. They colour the ten frames to record their work.

Name: ______________________ Date: ______________________

Counting by 10s

Circle groups of 10 ants.
Record the numbers.

________ groups of 10 and ________ left over ________ in all

Rouda used ten frames to organize her sticker collection.
How many stickers does she have?

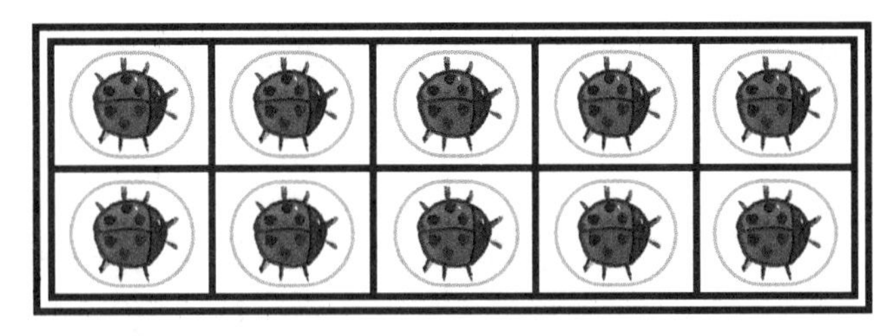

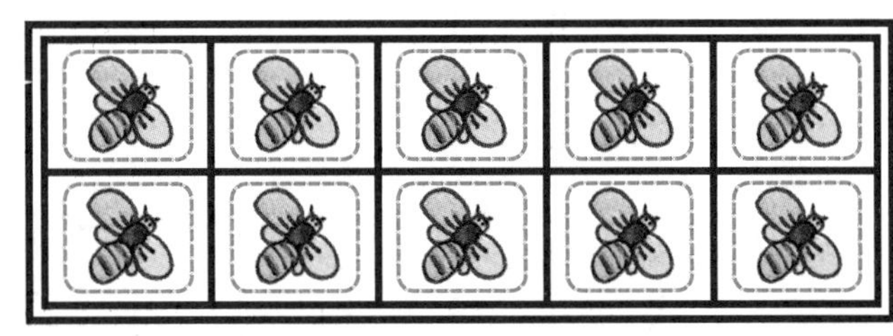

How can grouping by 10s help you with counting?

____

____

Focus | Children make groups of 10 and record the number of 10s, the leftover 1s, and the total.

Name: ____________________ Date: ____________________

Show Your Number

Count your objects.

How many do you have? ________

Draw a picture to show how many.

Groups of 10	1s (leftovers)

How many 10s? ________

How many 1s? ________

Write the number.

10s	1s

Focus | Children receive a set of cubes to count. They use a 2-part mat to show groups of 10s and 1s, then record their work.

Name: ______________________ Date: ______________________

63 Stars

Fill in the ten frames to show 63 stars.

There are ________ groups of 10 and ________ 1s.

Write the number.

10s	1s

HOME CONNECTION 

Select a number between 50 and 100. Ask your child to draw a picture for the number, then show you how many groups of 10, and how many 1s.

Focus | Children draw 63 stars in ten frames to show groups of 10s and 1s.

Name: ______________________ Date: ______________________

Same Number in Different Ways

Choose a 2-digit number.
Write the number. ________
Draw a place value mat for Tens and Ones.
Draw rods and cubes on the place value mat to show your number.

Show your number in a different way.
Use pictures, numbers, or words.

Focus | Children show a 2-digit number in different ways.

Name: ______________________ Date: ______________________

What's the Number?

Write the number of tens and ones. Then write the number.

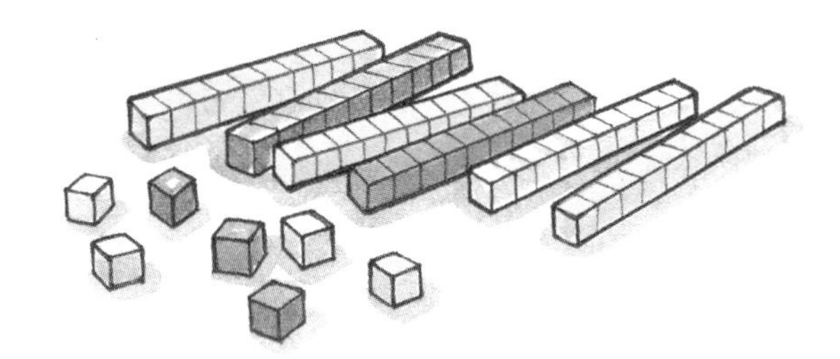

______ tens ______ ones

Write the number. ______

______ tens ______ ones

Write the number. ______

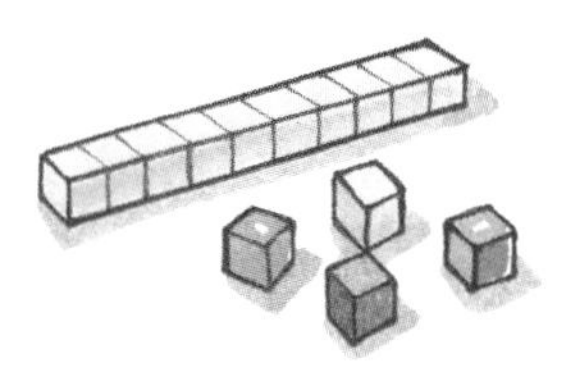

______ tens ______ ones

Write the number. ______

______ tens ______ ones

Write the number. ______

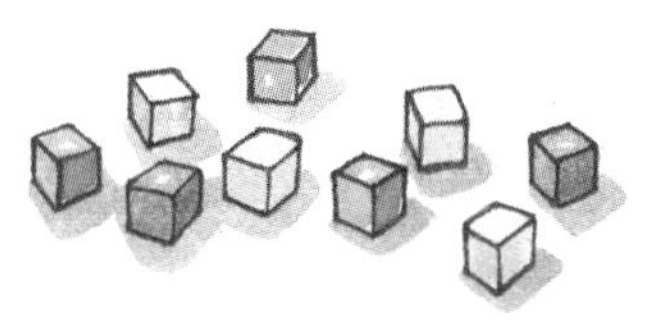

______ tens ______ ones

Write the number. ______

______ tens ______ ones

Write the number. ______

FOCUS | Children record numbers as tens and ones, and write the numbers.

Name: ______________________ Date: ______________________

Show the Coins

Show the number 56 with coins.

Show the number 56 with coins in a different way.

Show the number 93 with coins.

Focus | Children show 2-digit numbers with coins.

Name: ______________________ Date: ______________________

Which Coins Show 45¢?

You have 5 coins that equal 45¢.
What could the coins be? Draw a picture of your solution.

Focus | Children determine 5 coins that have a value of 45¢.

HOME CONNECTION 

Have your child explain the solution to the problem.
Ask: "How do you know the coins show 45¢?"

Name: ______________________ Date: ______________________

Which Coins Would You Use?

Which 6 coins would you use to make 55¢?
Use pictures, numbers, or words.

How do you know this makes 55¢?

____

Focus | Children determine which 6 coins they could use to make 55¢.

Name: ____________________ Date: ____________________

Changing Equal Sets

Use the 2 sets you are given.
Tell how you know they are equal.
Use pictures, numbers, or words.

Change the sets to make them unequal.
Draw the sets.

Tell how you made the sets unequal.
Use pictures, numbers, or words.

Focus | Children change 2 sets of identical objects that are equal in number to create unequal sets.

Name: ______________________ Date: ______________________

Making Unequal Sets

Use objects that are the same.
Construct 2 unequal sets.
How do you know the sets are unequal?

____

____

Draw your 2 unequal sets.

Focus | Children construct and draw unequal sets.

Name: ______________________ Date: ______________________

Which Set Is Not Equal?

Circle the set that is not equal in number to the other sets.

Explain your thinking using pictures, numbers, or words.

Focus | Children identify the set that is not equal in number to the other sets.

Name: ______________________ Date: ______________________

Be a Number Detective

Each piece of a 100-chart is missing some numbers.
Look for clues in the numbers to help you fill in the empty spaces.

22	23	24		26
	33	34	35	36
42		44	45	46

	52		54	55	
61	62	63		65	66
71	72		74		76

61		63	64	65
71	72	73		
	82		84	

	76		78	79	
	86	87	88		90
95	96		98		100

Page 47 fell out of a book.
How would you tell a friend where it belongs?

____

____

HOME CONNECTION 

Show a page number between 50 and 100 from a book. Ask: "What is the next page number? What was the number on the page that came before? How do you know?"

FOCUS | Children fill in the missing numbers on pieces of a 100-chart.

Name: ______________________ Date: ______________________

Missing Number Mysteries

Fill in some numbers in this part of a 100-chart.
Trade books with a friend.
Ask your friend to fill in the empty spaces.
Check your friend's work.

71									
									100

Choose 2 numbers in the 100-chart that are greater than 75.

Write the numbers. _____ _____

Compare the numbers. Use pictures, numbers, or words.

Focus | Children use a piece of a 100-chart to make a missing number puzzle for a friend to complete. Then, they check their friend's work.

Name: ______________________ Date: ______________________

Soccer Players

Order the numbers on the jerseys from least to greatest.

_______ _______ _______ _______

Tell how you know. Use pictures, numbers, or words.

Order the numbers on these jerseys from greatest to least.

_______ _______ _______ _______ _______

Tell how you know. Use pictures, numbers, or words.

FOCUS | Children order numbers from least to greatest and from greatest to least.

Name: ______________________ Date: ______________________

The Greatest and the Least

Choose 2 cards. Make the greatest number you can. _____
Show your thinking with pictures, numbers, or words.

Choose 2 cards again.
Make the least number you can. _____

Choose 2 cards again.
Make the least number you can. _____

Order the 3 numbers from least to greatest.

_____ _____ _____

Focus | Children choose cards to create 2-digit numbers, then order the numbers.

Name: ______________________ Date: ______________________

How Many Paddles?

There are 62 paddles altogether.
Show the number 62 in 3 ways.

Focus | Children show the number 62 in different ways.

Name: ______________________ Date: ______________________

My Journal

Tell what you learned about numbers to 100.
Use pictures, numbers, or words.

HOME CONNECTION 

Find out what your child learned about counting in this unit. Ask: "What is your favourite number? What are some different ways you can show it?"

Focus | Children reflect on and record what they learned about numbers to 100.

Addition and Subtraction to 18

Focus | Children create number stories about a scene from a country fair.

Name: ______________________ Date: ______________________

Dear Family,

In this unit, your child will be developing strategies for adding and subtracting 1-digit numbers.

The Learning Goals for this unit are to

- Develop and use mental math strategies to add and subtract two 1-digit numbers, including zero.
- Look for patterns in digits when adding and subtracting.
- Understand and use the relationship between addition and subtraction; for example, $2 + 7 = 9$ and $9 - 7 = 2$
- Find out if 2 sides of a number sentence are equal or not equal; for example, $10 = 6 + 4$ and $3 + 6 \neq 10$

You can help your child achieve these goals by doing the Home Connection activities suggested at the bottom of selected pages.

Name: ______________________ Date: ______________________

Number Stories from the Fair

Write 2 number sentences about each picture.

_____ + _____ = _____

_____ − _____ = _____

_____ + _____ = _____

_____ − _____ = _____

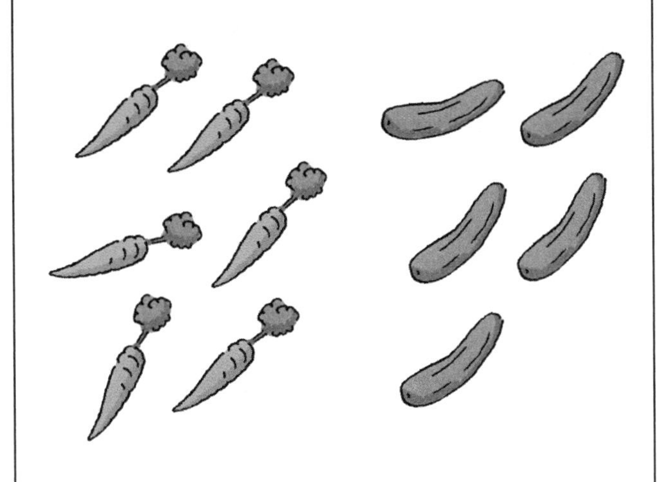

_____ + _____ = _____

_____ − _____ = _____

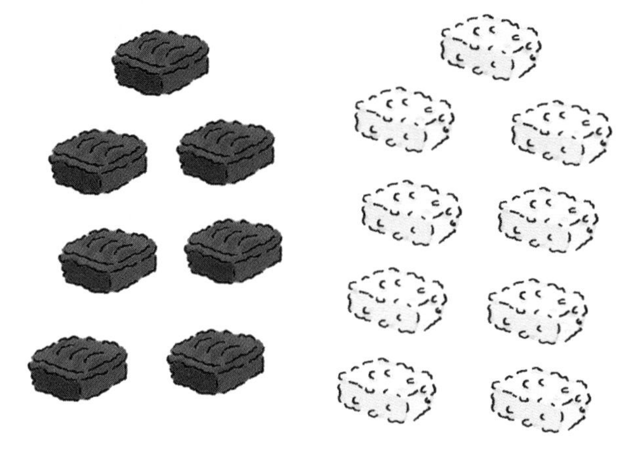

_____ + _____ = _____

_____ − _____ = _____

Focus | Children write addition and subtraction sentences about groups of items from a country fair.

Name: ______________________ Date: ______________________

Number Sentences

Write each number sentence.

_____ + _____ = _____

_____ + _____ = _____

_____ − _____ = _____

_____ − _____ = _____

Make your own.

_____ + 0 = _____

FOCUS | Children write number sentences about the pictures. They draw a picture to represent a number sentence.

Name: ______________________ Date: ______________________

Garden Problems

Write each number sentence.

There are 9  in a garden.

4 are yellow. The rest are pink.

How many  are pink? ____

____ ◯ ____ = ____

There were 13  on a tree.
Some fell off.

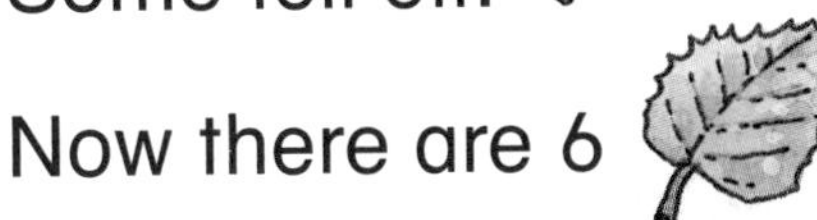

Now there are 6 on the tree.

How many fell off? ____

____ ◯ ____ = ____

There are 15  in the yard.

8  are in a tree.

The rest are on a fence.
How many are
on the fence? ____

____ ◯ ____ = ____

Make up your own story, then solve it.

Use the number 0 in your problem.

Focus Children write number sentences to represent addition- and subtraction-story problems.

HOME CONNECTION 

Share addition and subtraction story problems about things in your neighbourhood. For example, "There are 15 houses on our street. 9 of them have a garage. How many do not have a garage?"

Name: ______________________ Date: ______________________

Snappy Number Sentences

Use Snap Cubes. 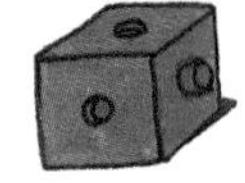

Write the number sentences.

_____ + _____ = _____ _____ − _____ = _____

_____ + _____ = _____ _____ − _____ = _____

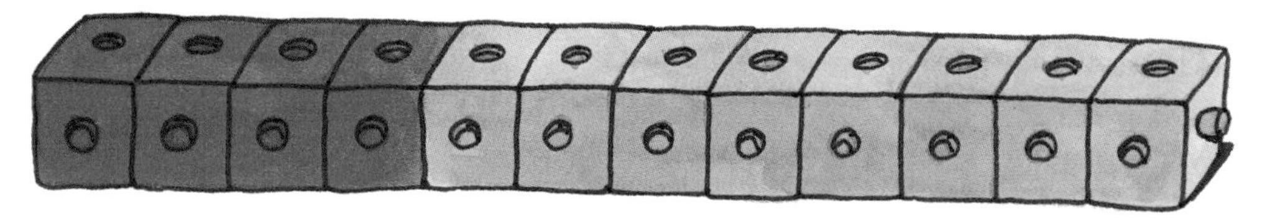

_____ + _____ = _____ _____ − _____ = _____

_____ + _____ = _____ _____ − _____ = _____

_____ + _____ = _____ _____ − _____ = _____

_____ + _____ = _____ _____ − _____ = _____

How does knowing 7 + 6 help when finding 13 − 6?

____

Focus Children write addition and subtraction sentences to describe arrangements of Snap Cubes. They explain how knowing an addition fact helps find the answer to a subtraction fact.

Name: ______________________ Date: ______________________

Add and Subtract

Write an addition sentence and a subtraction sentence for each picture.

___ + ___ = ___

___ − ___ = ___

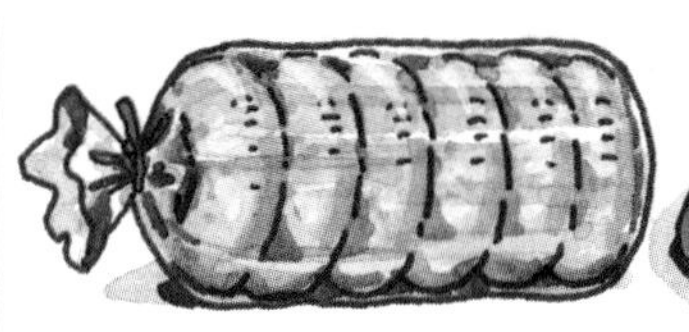

___ + ___ = ___

___ − ___ = ___

___ + ___ = ___

___ − ___ = ___

Draw your own picture.

3 + 8 = 11

11 − 3 = 8

Complete each number sentence.
Use counters to help or to check.

5 + 9 = ___	0 + 6 = ___	7 + 9 = ___
14 − 9 = ___	6 − 0 = ___	16 − 7 = ___

What patterns did you see? ______________________

____

Focus Children use the relationship between addition and subtraction to interpret picture stories.

HOME CONNECTION 

Have your child build a set of 5 to 9 pennies and then add 1, 2, or 3 pennies to that number. Have your child tell the addition sentence. Then, take the same number of pennies away. Have your child tell the subtraction sentence.

Name: ______________________ Date: ______________________

Write a Story Different Ways

Here are some number stories.
Tell each story 4 different ways.

____ + ____ = ____ ____ + ____ = ____ ____ − ____ = ____ ____ − ____ = ____	____ + ____ = ____ ____ + ____ = ____ ____ − ____ = ____ ____ − ____ = ____
____ + ____ = ____ ____ + ____ = ____ ____ − ____ = ____ ____ − ____ = ____	____ + ____ = ____ ____ + ____ = ____ ____ − ____ = ____ ____ − ____ = ____

How is subtraction like addition? ______________________

____

How is it different? ______________________

____

FOCUS | Children write addition and subtraction sentences to create fact families.

Name: ______________________ Date: ______________________

Secret Letter

Fill in the symbol that makes each sentence correct:

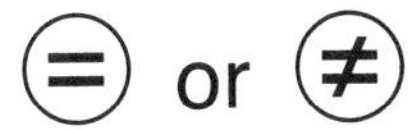

Use counters to help.

5 + 2 ◯ 6 + 1	6 + 7 ◯ 5 + 7	7 + 6 ◯ 13	1 − 0 ◯ 9 − 7
12 ◯ 7 + 5	14 − 9 ◯ 6	3 + 3 ◯ 9 − 3	6 − 5 ◯ 2
8 − 5 ◯ 3 + 0	1 + 7 ◯ 13 − 5	0 + 8 ◯ 4 + 4	16 − 8 ◯ 18 − 9
5 + 9 ◯ 8 + 6	17 − 8 ◯ 5 + 6	4 ◯ 12 − 8	17 − 8 ◯ 8
7 ◯ 8 − 1	3 + 6 ◯ 2 + 1	4 + 6 ◯ 9 + 1	7 + 5 ◯ 15 − 7

Colour the boxes where the symbol is =.

What is the secret letter? ______________________

Choose 1 sentence. Tell how you knew which sign to use.

____

HOME CONNECTION 

With your child, look around the home for items that come in equal sets. Discuss what happens if you add items or take a few items away. Are the sets still equal?

FOCUS | Children explore equal and unequal number sentences.

Name: ______________________ Date: ______________________

Yard Sale Day

The yard sale table is full.
Add each group of items in 2 different ways.

____ + ____ = ____ ____ + ____ = ____	____ + ____ = ____ ____ + ____ = ____
____ + ____ = ____ ____ + ____ = ____	____ + ____ = ____ ____ + ____ = ____
____ + ____ = ____ ____ + ____ = ____	What will you add to the yard sale? ____ + ____ = ____ ____ + ____ = ____

When might it help to change the order?

____

____

FOCUS | Children add numbers in different orders to recognize that the order does not matter.

Name: ______________________ Date: ______________________

Coin Count

How many pennies altogether?
How many ways can you add them?
Use pictures, numbers, or words.

Does the order matter when you add numbers? __________

How do you know? ______________________________

____

HOME CONNECTION 

When you set the table, have your child add groups of items on the table (forks, knives, glasses). Try adding them in a different order. Have your child explain why the order doesn't matter.

FOCUS | Children add the same numbers in different orders and explain that the order does not matter.

Name: ______________________ Date: ______________________

Bake Sale at the Fair

Some of the food at the bake sale is hidden.
Figure out what's missing.
Then write the addition sentence.

15
doughnuts

$\square + 4 = 12$

$7 + \square =$ ____

16
chocolates

$2 + \square =$ ____

$\square + \square = 16$

Choose 1 addition sentence. Tell how you solved it.

HOME CONNECTION 

Have your child take a number of cans from the cupboard. Use a cloth to hide some of them. Have your child explain how addition can be used to find out how many are covered.

Focus | Children explore using addition to solve missing addend problems.

Name: ______________________ Date: ______________________

How Many More Dogs?

There are 8 dogs ready to
compete in the dog show.
17 dogs were signed up for the show.
How many more dogs have to come?

Tell how you solved the problem.
Use pictures, numbers, or words.

Make up your own "missing parts" story.
Show how you solved it.

HOME CONNECTION 

When you spend leisure time with your child, pose different missing parts problems. Have your child explain how he or she figured out the answer.

FOCUS | Children solve missing parts problems using their own strategies.

Name: ______________________ Date: ______________________

How Does the Story Start?

Jia Li went down 4 floors on the elevator.
She got out on Floor 7.
Where did Jia Li start?
Use addition to help you.

$\square - 4 = 7 \qquad 7 + 4 = 11$

Make up your own elevator problem and solve it.
Use pictures, numbers, or words.

Leo spent $6 at the yard sale.
He has $8 left.
How much did he start with?

$\square - 6 = 8 \qquad$ ____ + ____ = ____

Mahala gave Fido 1 treat each day for 1 week.
She has 3 treats left.
How many treats did Mahala buy?

$\square - 7 = 3 \qquad$ ____ + ____ = ____

Focus | Children solve problems by completing subtraction sentences with missing numbers.

Name: ______________________ Date: ______________________

What's the Difference?

Drop 1 counter on each game board.

15	10	12	17
11	16	10	12
18	12	15	14
17	16	13	18

4	7	8	3
1	0	5	2
9	3	5	4
0	6	3	9

Write 1 number in each box.

Complete each subtraction sentence.

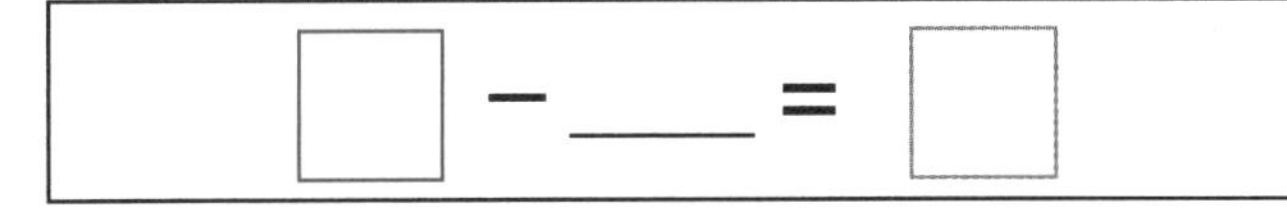

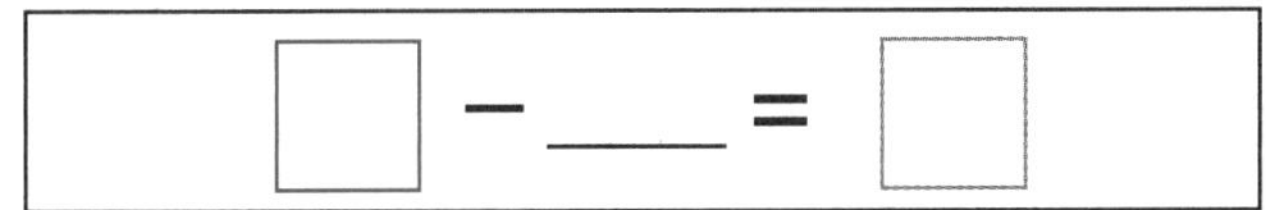

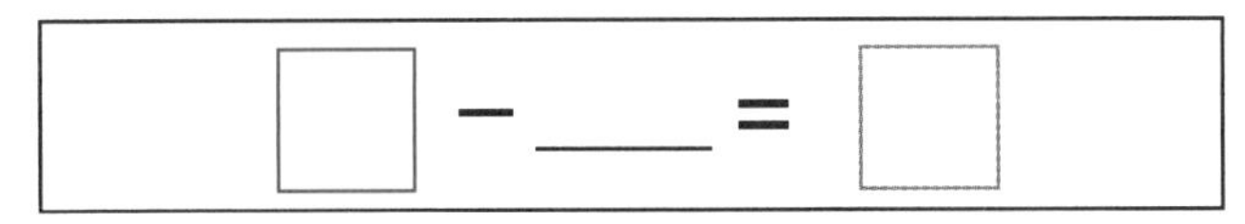

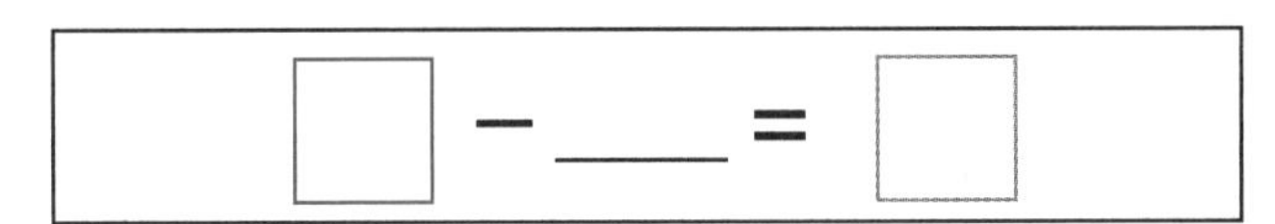

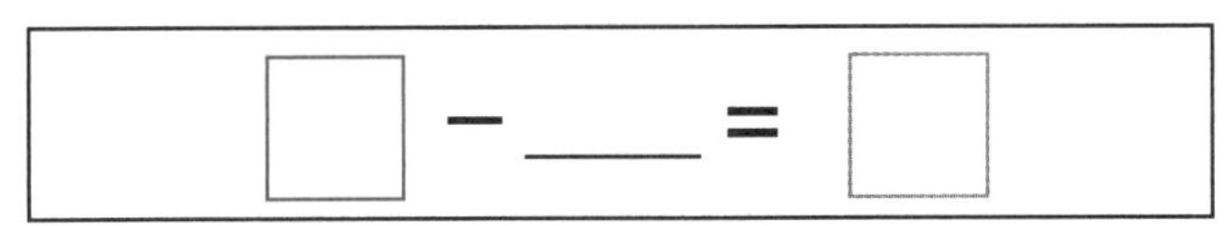

☐ − ___ = ☐

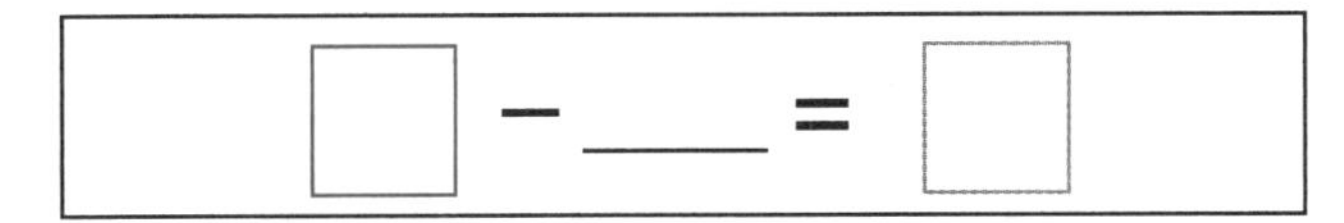

☐ − ___ = ☐

Choose 1 subtraction sentence. Tell how you solved it.

____

____

Focus | Children find the missing number in a subtraction sentence.

Name: ____________________ Date: ____________________

Find the Missing Number

Complete each number sentence.
Then answer the question.

7 ◯ ____ = 15

15 ◯ ____ = 7

15 ◯ 7 = ____

How many beads will go on the bracelet? ____

12 ◯ ____ = 8

____ ◯ 8 = 12

12 ◯ 8 = ____

How many eggs are missing? ____

____ ◯ ____ = ____

____ ◯ ____ = ____

____ ◯ ____ = ____

How many children have not come yet? ____

HOME CONNECTION 

When you solve everyday subtraction problems, ask your child questions such as, "What do we already know?" "What piece is missing?"

Focus | Children write addition and subtraction sentences that describe picture stories.

Name: ______________________ Date: ______________________

Baseball Subtraction

Write a subtraction sentence for each picture.

___ – ___ = ___

___ – ___ = ___

___ – ___ = ___

___ – ___ = ___

How did you decide which number to start with? ______________________

____

____

Focus | Children recognize that order matters in subtraction.

Name: ______________________ Date: ______________________

Any Room on the Rocket Blaster?

There are 18 empty seats on the Rocket Blaster.
Some people get on.

Now there are 9 empty seats.
How many people got on?

Tell how you solved the problem.
Use pictures, numbers, or words.

Make up your own story about subtracting 0.

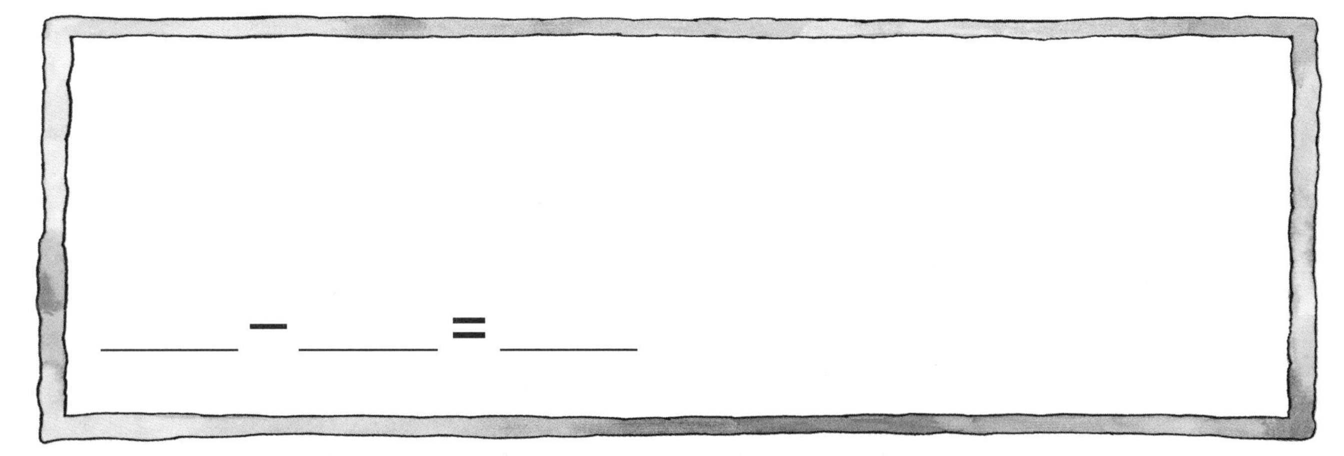

_____ − _____ = _____

Focus | Children solve subtraction stories using their own strategies.

Name: ______________________ Date: ______________________

Our Class Loves Pets!

Indira compared the numbers of pets.
Which pets is Indira comparing? ______________

Draw a picture to explain.

Write the number sentence. ___ ◯ ___ = ___

Compare your answer with those of your friends.
Write a number sentence for each different answer.

___ ◯ ___ = ___

___ ◯ ___ = ___

___ ◯ ___ = ___

Focus | Children solve a problem that has several possible answers.

Name: ______________________ Date: ______________________

I or 2 More

How many buttons?
Circle the number.

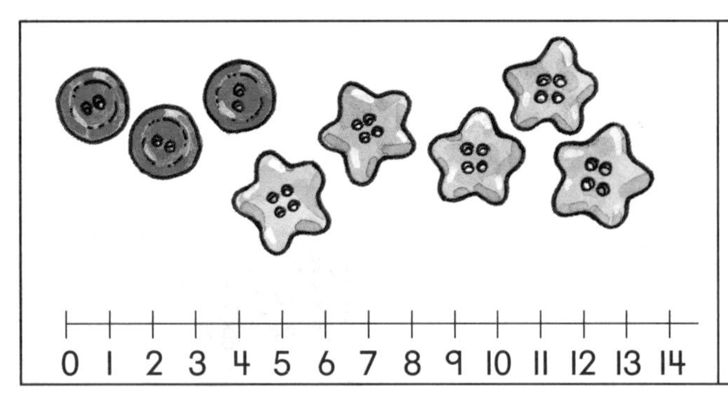

0 1 2 3 4 5 6 7 8 9 10 11 12 13 14

Add 1 more .

___ + ___ = ___

Circle the sum.

0 1 2 3 4 5 6 7 8 9 10 11 12 13 14

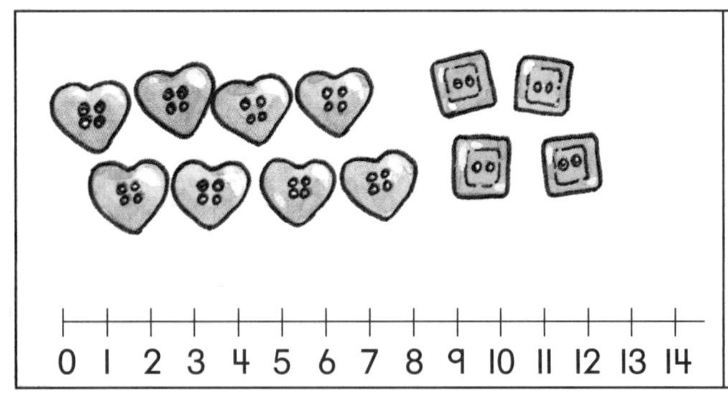

0 1 2 3 4 5 6 7 8 9 10 11 12 13 14

Add 2 more .

___ + ___ = ___

Circle the sum.

0 1 2 3 4 5 6 7 8 9 10 11 12 13 14

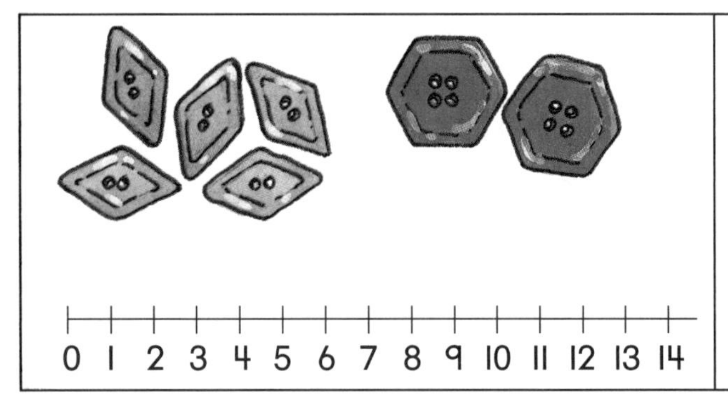

0 1 2 3 4 5 6 7 8 9 10 11 12 13 14

Add 2 more .

___ + ___ = ___

Circle the sum.

0 1 2 3 4 5 6 7 8 9 10 11 12 13 14

How did you add 1 more? ______________________

How did you add 2 more? ______________________

Focus | Children add 1 or 2 more to a number.

Name: ______________________ Date: ______________________

1 or 2 Less

How many buttons?
Circle the number.

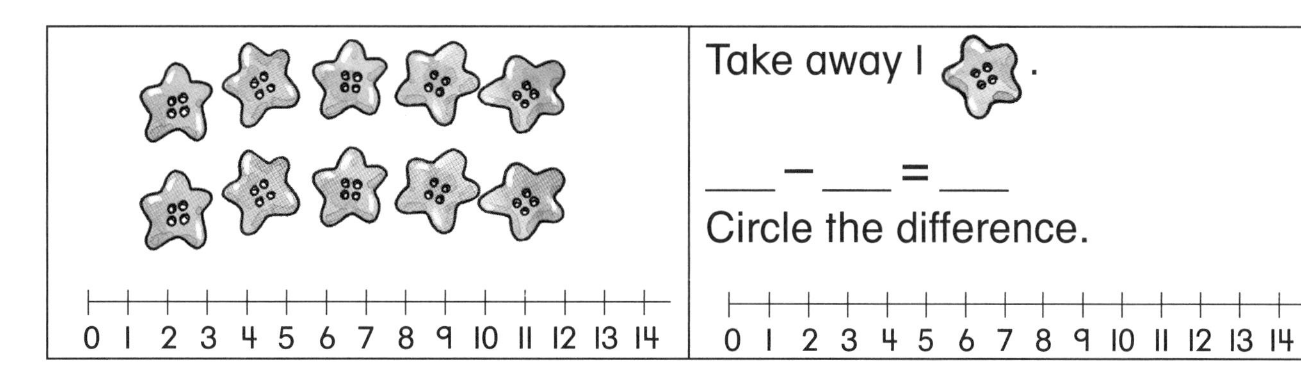

0 1 2 3 4 5 6 7 8 9 10 11 12 13 14

Take away 1 .

___ – ___ = ___

Circle the difference.

0 1 2 3 4 5 6 7 8 9 10 11 12 13 14

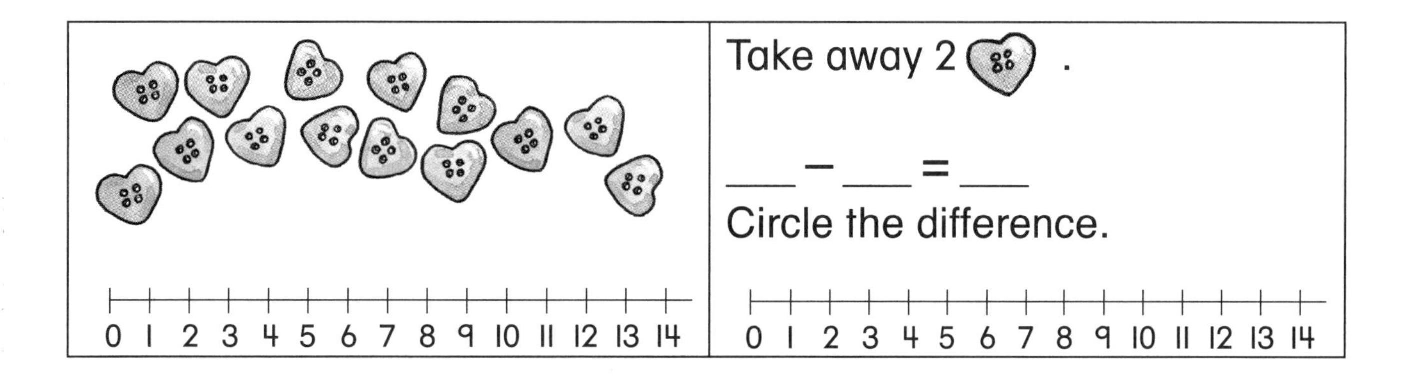

0 1 2 3 4 5 6 7 8 9 10 11 12 13 14

Take away 2 .

___ – ___ = ___

Circle the difference.

0 1 2 3 4 5 6 7 8 9 10 11 12 13 14

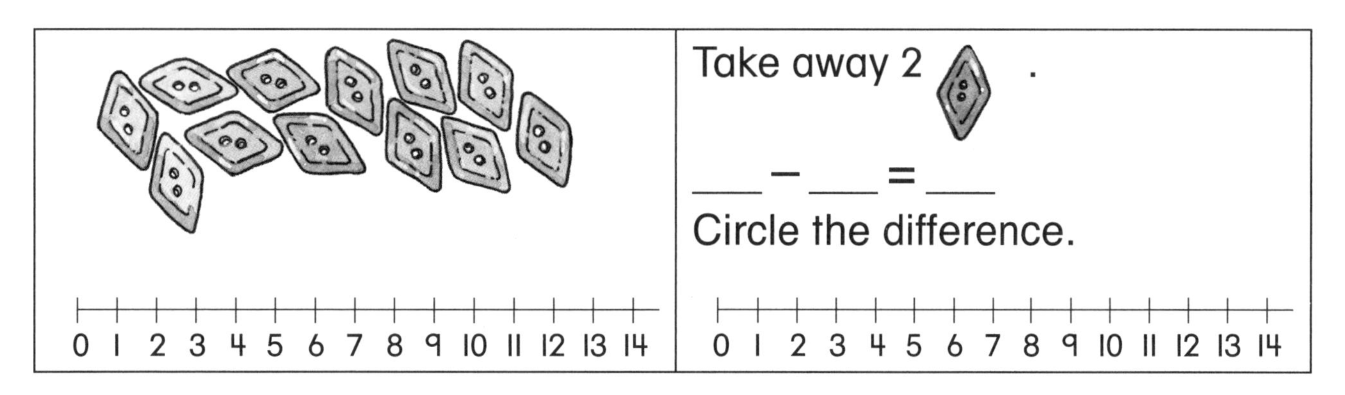

0 1 2 3 4 5 6 7 8 9 10 11 12 13 14

Take away 2 .

___ – ___ = ___

Circle the difference.

0 1 2 3 4 5 6 7 8 9 10 11 12 13 14

How did you take away 1? ______________________

How did you take away 2? ______________________

Focus | Children subtract 1 or 2 from a number.

Name: ______________________ Date: ______________________

Adding Rows and Columns

Look at the charts of numbers.
Add the numbers in each row.
Add the numbers in each column.

column

row →	9	5	____
row →	8	9	____
	____	____	

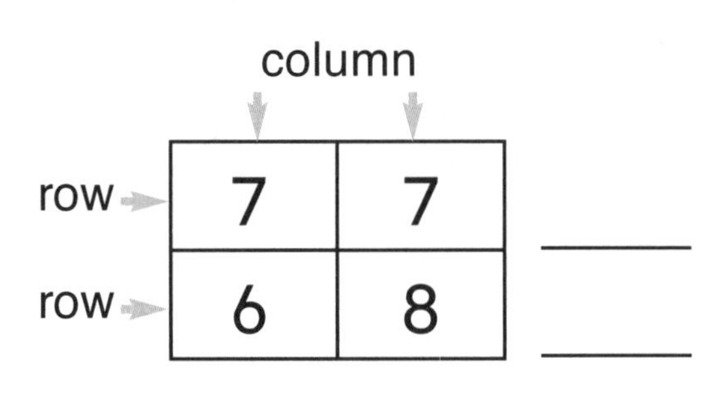

When did the "make 10" strategy help you?

____

____

When did the "make 10" strategy *not* help you?

____

____

What other strategies did you use?

____

____

____

Focus | Children use different strategies to add 2 numbers.

HOME CONNECTION 

Use 2 digits to write and solve addition sentences with your child. For example, if the digits are 8 and 9, the sentence is 8 + 9 = 17.

Name: ______________________ Date: ______________________

Finding Doubles

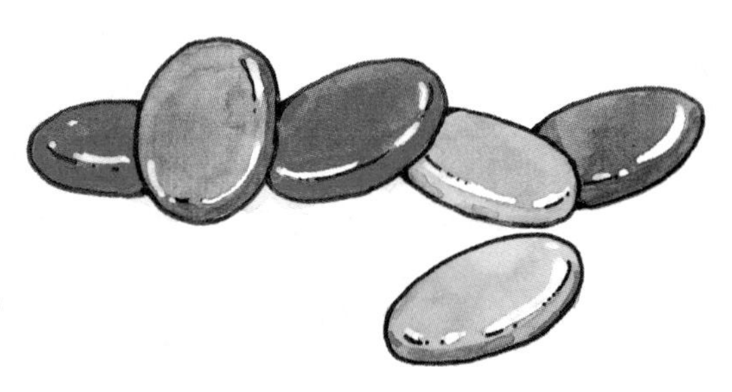

Use counters.
Which sums can be written as doubles?

8 + 10   2 + 9   0 + 2   5 + 7   9 + 7

3 + 4   5 + 6   2 + 3   3 + 6   0 + 1   6 + 7

9 + 6   8 + 6   5 + 8   1 + 3   2 + 3

Draw the counters.
Write the addition sentences.

9 + 7 = ____ + ____

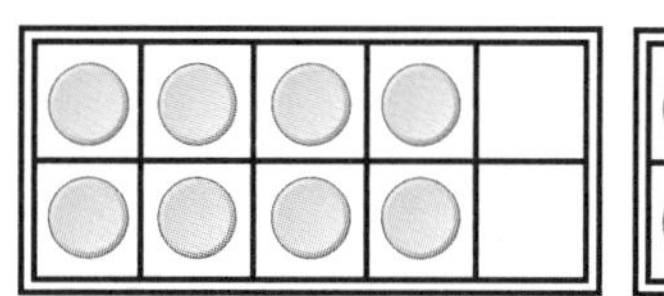
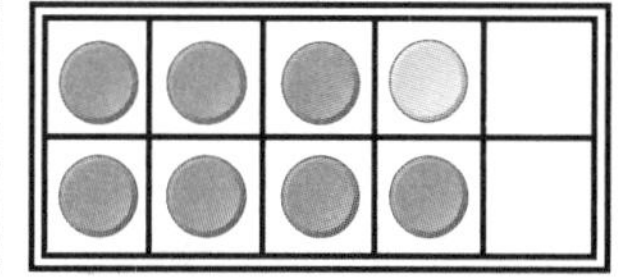

____ + ____ = ____ + ____

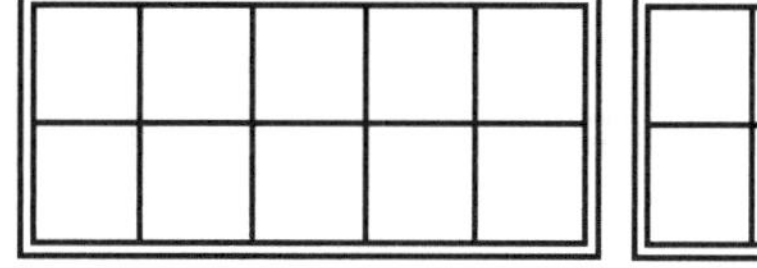
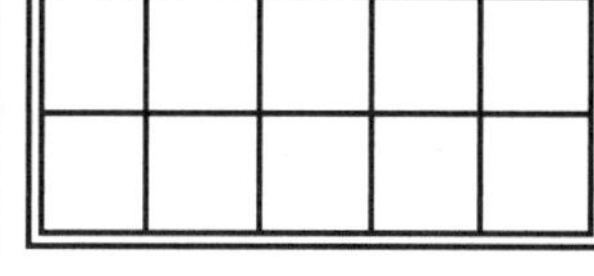

____ + ____ = ____ + ____

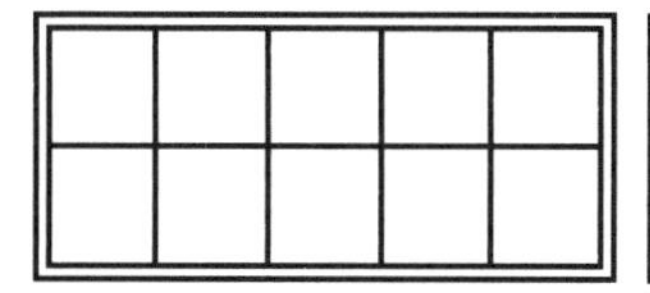
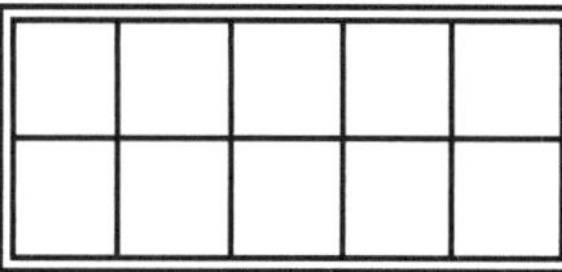

____ + ____ = ____ + ____

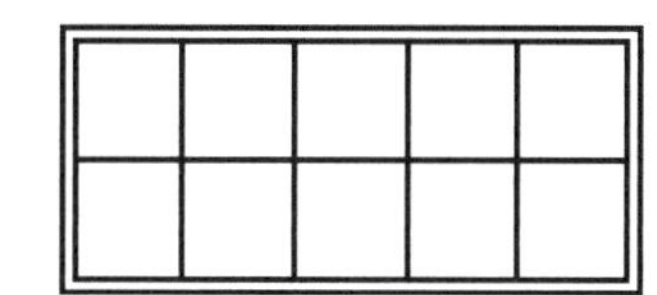
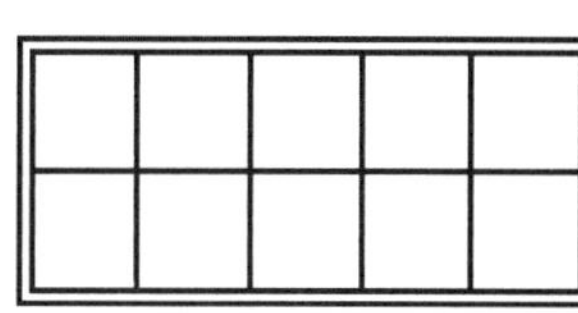

Focus | Children investigate which addition facts can be turned into doubles facts.

Name: ______________________ Date: ______________________

Seeing Doubles

Write 2 doubles facts that can help you find each answer.

6 + 7 = ___	5 + 6 = ___	8 + 7 = ___
__ + __ + __ = __	__ + __ + __ = __	__ + __ + __ = __
__ + __ − __ = __	__ + __ − __ = __	__ + __ − __ = __
7 + 9 = ___	5 + 3 = ___	6 + 8 = ___
__ + __ + __ = __	__ + __ + __ = __	__ + __ + __ = __
__ + __ − __ = __	__ + __ − __ = __	__ + __ − __ = __

HOME CONNECTION 

Ask your child to tell a number story using a double or near doubles.

Focus | Children use doubles facts to find answers to near doubles.

Name: ______________________ Date: ______________________

Doubles Patterns

Write sums for Patti's facts. Colour the boxes red.
Write sums for Joe's facts. Colour the boxes blue.

+	0	1	2	3	4	5	6	7	8	9
0										
1										
2										
3										
4										
5										
6										
7										
8										
9										

What patterns do you notice?

Focus | Children use doubles and near doubles to complete an addition table.

Name: ______________________ Date: ______________________

Finding Doubles

Here are sums of 2 numbers.

1	2	3	4	5	6	7	8	9
10	11	12	13	14	15	16	17	18

Which sums can you write as doubles?
Use pictures, numbers, or words to explain.

How can you tell which sums you can write as doubles and 1 more?

How can you tell which sums you can write as doubles and 2 more?

Focus | Children identify sums as doubles or near doubles facts.

Name: ______________________ Date: ______________________

Adding in Different Ways

Add the numbers 2 different ways.

6 7 3 2	___ + ___ + ___ + ___ = ___ ___ + ___ + ___ + ___ = ___
1 5 4 8	___ + ___ + ___ + ___ = ___ ___ + ___ + ___ + ___ = ___
9 0 4 5	___ + ___ + ___ + ___ = ___ ___ + ___ + ___ + ___ = ___
3 5 5 3	___ + ___ + ___ + ___ = ___ ___ + ___ + ___ + ___ = ___

Choose 1 set of cards. Which way is easier to add?

____

Why is the sum the same each time?

____

Focus | Children combine more than 2 addends in different ways.

Name: ______________________ Date: ______________________

What's the Story?

The sum is 16.
What could the story be?
Use pictures, numbers, or words.

_____ + _____ = 16   _____ + _____ = 16

The difference is 3.
What could the story be?
Use pictures, numbers, or words.

_____ − _____ = 3   _____ − _____ = 3

Tell about adding 0. ______________________

____

____

Tell about subtracting 0. ______________________

____

____

Focus | Children show their understanding of addition and subtraction by completing number sentences.

Name: ______________________ Date: ______________________

Add and Subtract

Make 2 sets of counters like these:
Move your counters to change the numbers in your sets.
Draw pictures and write number sentences.

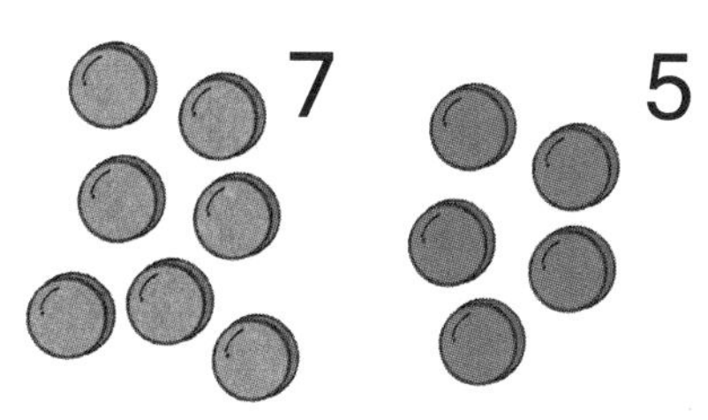

$7 + 5 =$ ____ $+$ ____	$7 + 5 =$ ____ $+$ ____

Grandma added fish to each tank. How many fish did she add?

____ $- 6 =$ ____

____ $+$ ____ $= 13$

____ $- 4 =$ ____

____ $+$ ____ $= 11$

FOCUS | Children relate addition and subtraction.

Name: ______________________ Date: ______________________

My Journal

Which strategy do you use most often?

Tell a story about a time when a strategy helped you to add or subtract.

Use pictures, numbers, or words.

HOME CONNECTION 

Watch for situations where your child can use a strategy to add or subtract 1-digit numbers.

Focus | Children reflect on the usefulness of addition and subtraction strategies.

The Skating Day

The class looked outside at the ice and the snow.
"If this storm doesn't stop soon, I don't think we'll go."
Then Cam's grandma came. "All the roads are okay.
We'll still take the bus to go skating today."

It was so frosty cold when they got on the bus,
they climbed on it quickly without any fuss.
Miss Chu called out names, as they passed her in line.
She didn't want any children left behind.

The volunteers helped them lace their skates on tight.
It was lots of hard work to tie them just right.
"It's freezing!" said Mishi. Her teeth loudly chattered.
"I love skating!" cried Cam. To him, nothing else mattered.

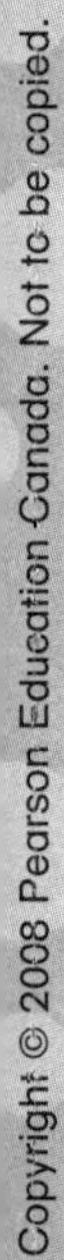

Once on the ice, it was easy to spot
the children who skated and those who did not.
Grandma flew by; she was spinning and turning.
She encouraged the children: “Once I was just learning.”

Then Grandma stopped skating. “My scarf! Where’s it gone?
I’m sure that this morning I put the scarf on.”
They looked on the rink, but the scarf wasn’t there.
They looked by the benches; they looked everywhere.

“Who’d want my scarf? That is puzzling to me.”
Cam tugged Grandma’s coat. Mishi said, “It was me.
I felt freezing cold, so I wrapped myself in it.
I thought I would borrow it just for a little bit.”

Grandma smiled. “I’m so glad that the scarf has not gone.
It’s my favourite for winter, but you keep it on.
Let’s skate a while longer, before leaving the rink.
Then, to warm up, we’ll have hot chocolate to drink!”

About the Story

The story was read in class to prepare for a Mathematics Investigation activity. Children completed addition and subtraction activities, worked with number combinations, and counted in a variety of ways. They also identified and made their own patterns and used coins to represent different money amounts.

Talk about It Together

- What happens to Grandma's scarf while the class is at the skating arena?
- How do you think the parent volunteers might be using math during the field trip?
- Why did Mishi borrow the scarf?
- What kind of person is Grandma? Do you know someone like her? How is that person the same? Different?

At the Library

Ask your local librarian about other good books to share about patterning, numbers, and counting money.

How Many Children?

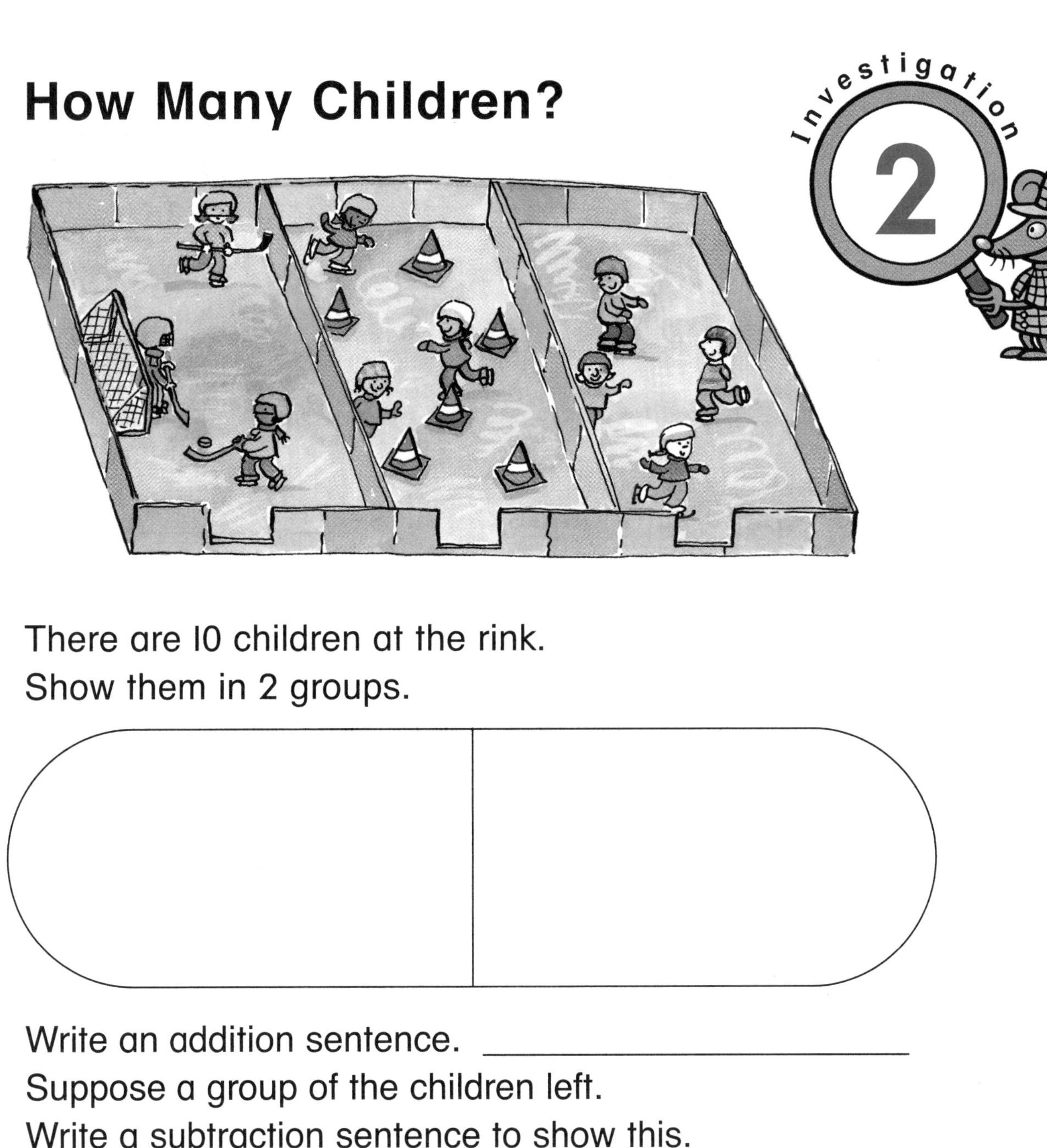

There are 10 children at the rink.
Show them in 2 groups.

Write an addition sentence. ______________________

Suppose a group of the children left.

Write a subtraction sentence to show this. __________

Show the 10 children in 2 groups in a different way.

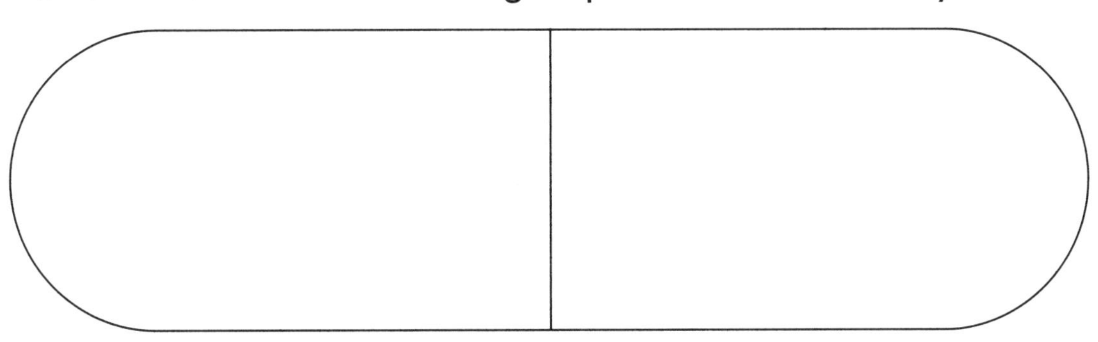

Write an addition sentence and a subtraction sentence.

____

Buy Cam a Snack!

Grandma has this money in her purse.

What is the value of Grandma's pennies? __________

What is the value of Grandma's nickels? __________

What is the value of Grandma's dimes? __________

Grandma spends 86¢
on Cam's snack.
What coins might Grandma
use to buy Cam's snack?
Use pictures, numbers, or words.

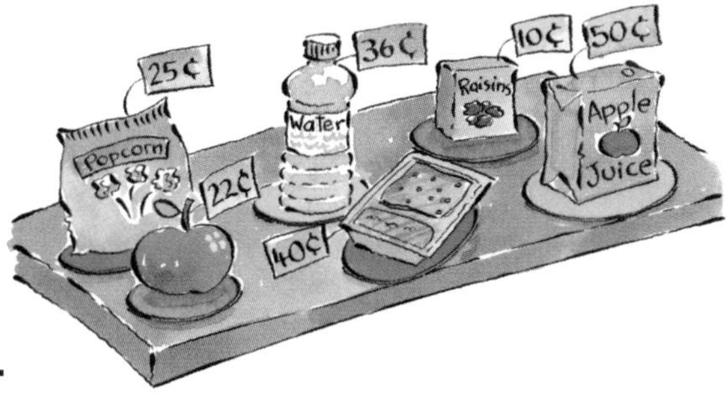

Show another way to make 86¢.
Use pictures, numbers, or words.
Show 1 or more quarters.

Grandma's Scarf

What is the rule for the repeating pattern
on Grandma's scarf?____________________________

____

Show the pattern, using 3 repeats.
Mark the pattern core.

Make an increasing pattern for a scarf.
Draw your pattern.

Describe your increasing pattern.

____

____

Race to 100!

Roll 2 number cubes.
Add the numbers.
Take that number of Snap Cubes.

Every time you have 10 Snap Cubes,
put them together to make a 10-stick.
Stop when you have 10 sticks. You will have 100 cubes!

Estimate. How many rolls will you need to get 100 cubes?
Show your thinking in pictures, numbers, or words.

Keep track. Make a tally for each roll.
When you reach 100, count the tallies.

How many rolls altogether? ________

Floating Bubbles

Challenge a friend to see
whose bubble stays
in the air longer.
Blow at the same time
and begin counting slowly.

Do you think
you will get to 10? 20?
Will you get as high
as 50?

Food for Thought

Fill in the blanks with something that makes sense.

I could eat 100 ____________________

but not 100 ____________________.

I could lift 100 ____________________

but not 100 ____________________.

I would like to have 100 ____________________

but not 100 ____________________.

Make up some more sentences of your own.

 The next 4 pages fold in half to make an 8-page booklet.

Fold

Math at Home

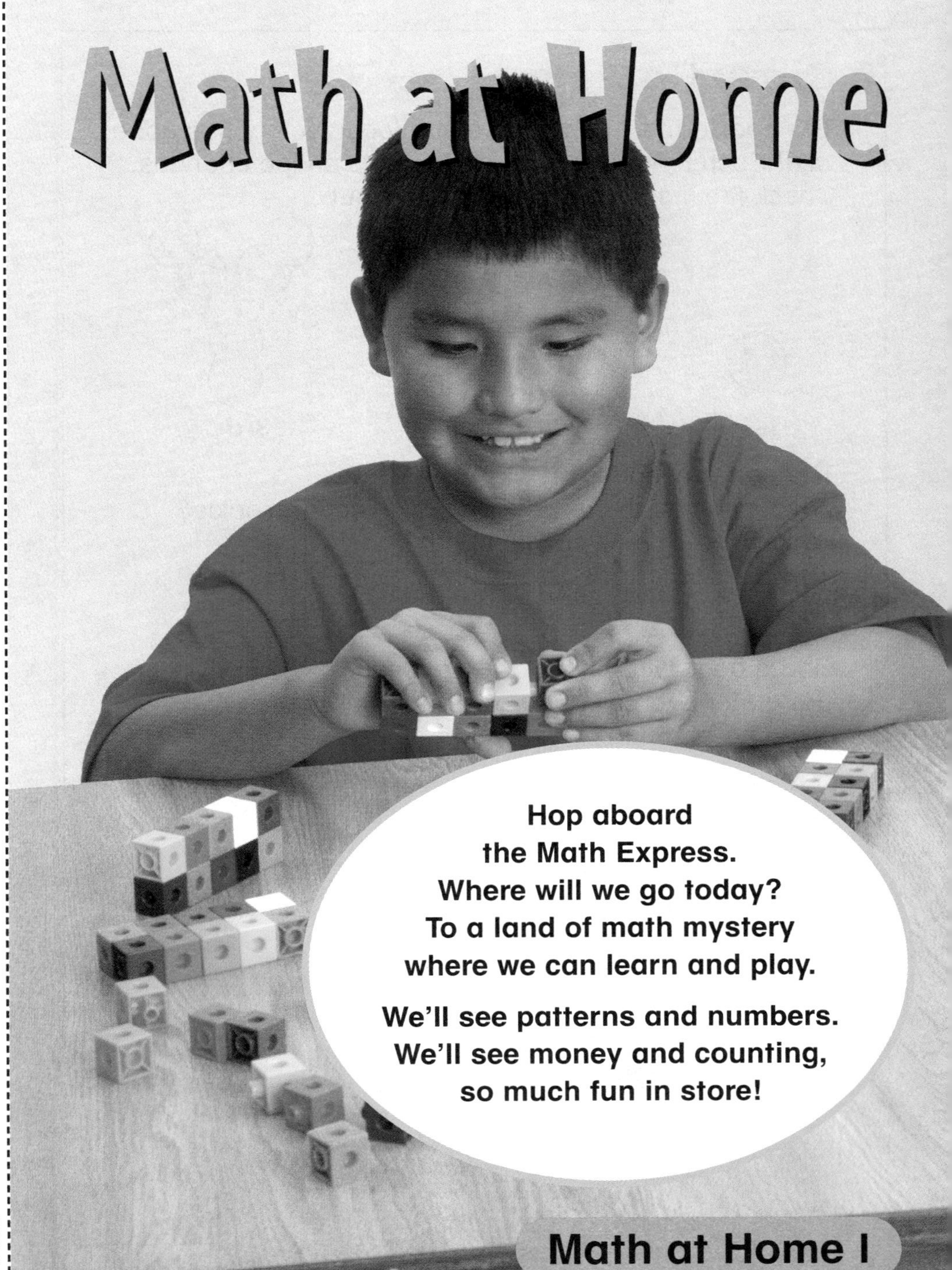

Math at Home 1

Crazy Cookies

The cookie-making machine at a local factory has gone wild! Each time a cookie pops out, its shape changes. Check the first 3 cookies that came out.

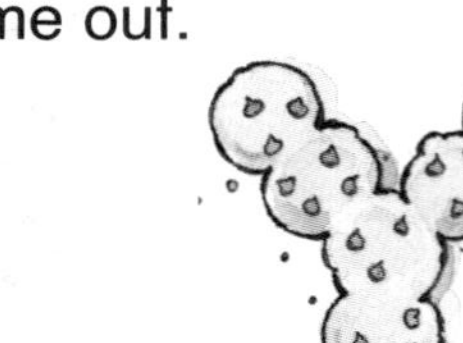

Ist **2nd** **3rd**

The pattern continues.
What will the 5th cookie look like? The 7th cookie?
Which one would you like to eat?

Find Your Page!

Find a book with almost 100 pages in it.
Get a friend to call out a page number that could be in the book.

Open the book as close as you can to that page number.
Estimate how far off you were, then give your friend a turn.

What was the closest you got?
Would it be easier if the book had 50 pages? Why?

Pattern Search

At home, look for patterns made with
- colours
- pictures
- different objects
- numbers

Can you find other types of patterns?

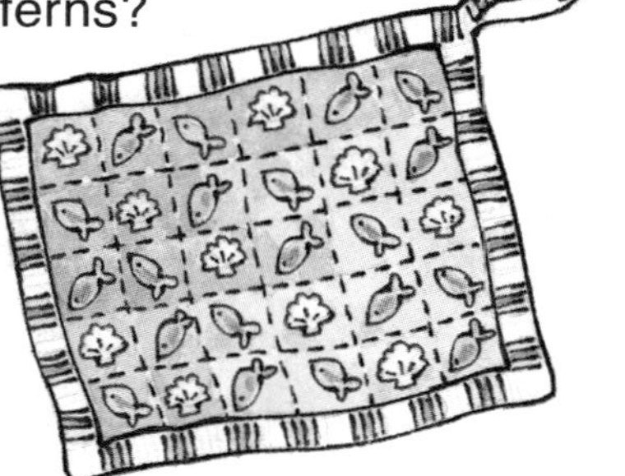

Elevating Elevators

Imagine you are in a tall building.
You leave the doctor's office and get on the elevator at floor 11.
You need to go down 5 floors to get to the cafeteria.

Which button will you push?

Suppose you got on at floor 3 and went up 9 floors.
On which floor will you get off?

Make up some elevator problems of your own.

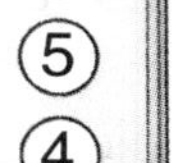
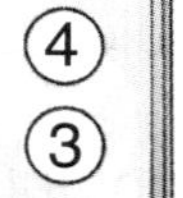
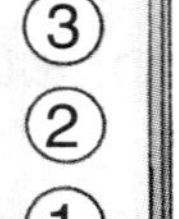

Doubles Hunt

At the grocery store, look to see how many items you can find that are packaged in doubles. When you find one, figure out how the numbers would change if you added a few more or a few less.

Here are some items to get you started:

What others can you find?

Penny Problems?

Put 13 pennies in a cup.
Spill them out.

Record a number sentence that tells the number of heads and the number of tails and the sum.

If you do it again, will it be the same?
How many ways are there? Try it and see!

7 + 6 = 13

Dirty Laundry

Max ran through a big puddle. He shook off the mud beside the clean laundry.

Estimate how many mud spots landed on the sheet. (Think groups of 10, then the estimating will be a breeze!)

Adding Stars

How many stars altogether?

Sean said, "8 and 2 more is 10. And 4 more is 14."

Amy said, "I know 8 and 8 is 16.
If I take 2 away, it's 14."

Who's right? Why do you think so?

Secret Numbers

You'll need:
- 5 sets of number cards from 0 to 9, shuffled and placed face down
- ten frames
- 20 counters

On your turn:
- Draw 2 cards and place them **face up** in front of you. These cards show 73.
- Draw 2 more cards and place them **face down** in front of a friend.

Your friend has a choice:
- Trade a card for one of yours.
- Leave the cards alone.

Once you trade or keep cards, look at the numbers.
- Flip over the face-down cards, then read both numbers.
- The player with the greater number puts a counter on his ten frame.
- The cards go into the discard pile. The other player draws the next cards.

The first player to fill a ten frame wins!

Ten Frames for Secret Numbers

Can you think of a different game to play with the same materials?

Measurement

Focus | Children look for objects that can be compared or ordered by length or mass.

Name: ______________________ Date: ______________________

Dear Family,

In this unit, your child will be learning about measurement. Your child will learn to read a calendar and develop an understanding of linear measurement, distance around, and mass.

The Learning Goals for this unit are to

- Learn about days in a week and months in a year.
- Read a calendar to find the date.
- Estimate, measure, and compare lengths using non-standard units, such as paper clips, straws, or Snap Cubes.
- Estimate, measure, and compare the distance around an object using non-standard units.
- Estimate, measure, and compare masses using balance scales and non-standard units such as beads, blocks, or Snap Cubes.
- Solve everyday problems about measurement.

You can help your child achieve these goals by doing the Home Connection activities suggested at the bottom of selected pages.

Name: ______________________ Date: ______________________

Comparing Objects

Find 2 objects and compare the lengths or heights.

I compared ____________ and ____________ .

Which is shorter? ____________

Tell how you compared. Use pictures, numbers or words.

Find 2 objects and compare the masses.

I compared ____________ and ____________ .

Which is heavier? ____________

Tell how you compared. Use pictures, numbers or words.

FOCUS | Children choose objects and compare the lengths or heights. They choose 2 other objects and compare the masses.

HOME CONNECTION 

Have your child compare and order objects in your home using words such as longest/shortest and heaviest/lightest. Ask your child to explain how he or she compared them.

Name: ____________________ Date: ____________________

Eddie's Calendar

Fill in the missing days of the week on Eddie's calendar.

October

	Monday				Friday	
		1	2	3	4	5 Dentist
6	7	8	9 My Birthday	10	11	12
13 Visit Grandma	14	15	16	17 Swimming with Tom	18	19
20	21	22	23	24	25	26
27	28	29	30	31		

Today is Eddie's birthday. What day of the week is it? ____________

What day was yesterday? ____________

What day of the week did Eddie go to the dentist? ____________

What day will he go swimming? ____________

What day of the week is October 26? ____________

Focus | Children identify the day of the week for today, yesterday, and tomorrow, and for a given date.

HOME CONNECTION 

Make a weekly calendar with your child. Talk about the family schedule and what happens on each day of the week. Have your child identify the day of the week and the date tomorrow and yesterday.

Name: ______________________ Date: ______________________

Reading a Calendar

Use Eddie's calendar on page 108.
Today Eddie is going swimming.

What is the date tomorrow? ____________

What was the date yesterday? ____________

What is the date in 1 week? ____________

Has it been more or less than a week

since Eddie saw the dentist? ____________

Use pictures, numbers, or words to tell how you know.

Eddie leaves for holidays in 1 week and 4 days.

What date will he leave? ____________

Use pictures, numbers, or words to tell how you know.

HOME CONNECTION 

Mark a calendar for the current month with your child to show dates of any holidays and special events.

Focus | Children use a calendar to solve problems involving the number of days in a week.

Name: ______________________ Date: ______________________

Days and Weeks

Complete the calendar.

June

		Tuesday		Thursday		Saturday
			1	2	3	4
5	6	7	8	9	10	11
12	13	14	15	16	17	18
19	20	21	22	23	24	25
26	27	28	29	30		

What day of the week is June 15? ____________

What day of the week is 4 days before Friday? ____________

What date comes 2 days after June 21? ____________

What is the date of the third Sunday in June? ____________

How many days is it from June 9 to June 17? ____________

Is that more or less than a week?

Use pictures, numbers, or words to tell how you know.

Focus | Children read a calendar and identify the dates in a month and solve related problems.

Name: ______________________ Date: ______________________

Months of the Year

Print the months in order in the chart.

September	January	May	July
February	August	December	October
March	April	November	June

1.	7.
2.	8.
3.	9.
4.	10.
5.	11.
6.	12.

What is the sixth month? ____________

What month is after January? ____________

What month is 2 months after May? ____________

What month is 5 months after October? ____________

What month is 3 months before June? ____________

How many months are in a year? ____________

HOME CONNECTION 

Make a chart of the months of the year and record the birthdays of family members and friends.

FOCUS | Children order the months of the year and solve related problems.

Name: ______________________ Date: ______________________

Which Is Longest?

Find 3 objects to measure.

I chose ______________, ______________,

and ______________.

What unit did you use? ______________

Show how you measured.
Use pictures, numbers, or words.

Order the lengths from shortest to longest.

______________ ______________ ______________

Tell how you ordered the lengths.
Use pictures, numbers, or words.

HOME CONNECTION

Choose 3 objects in your home that you and your child can measure and order by length (for example, kitchen utensils, sports equipment) using units such as paper clips or unsharpened pencils.

Focus | Children measure, compare, and order the lengths of objects.

Name: ______________________ Date: ______________________

Choosing a Unit

Choose a unit to use to measure 3 lengths. ______________
Estimate before each measure.

What I measured	My estimate	My measurement
	about ________	
	about ________	
	about ________	

Choose another unit. ______________
Estimate and measure the same 3 lengths again.

What I measured	My estimate	My measurement
	about ________	
	about ________	
	about ________	

Choose a length you measured.
How does the unit change the measurement?

____

____

What unit would you choose next time? Tell why.

____

____

Focus | Children choose units to estimate, measure, and order lengths.

HOME CONNECTION 
Ask your child to explain why you need to use the same units for each length when ordering lengths.

Name: ______________________ Date: ______________________

Over and Over

Use only 1 copy of a unit to measure lengths.
Measure 2 lengths that are straight lines.

What I measured	The unit I chose	My estimate	My measurement
		about ______	
		about ______	

Measure 2 lengths that are not straight lines.

What I measured	The unit I chose	My estimate	My measurement
		about ______	
		about ______	

Choose an object. ____________ Choose a unit. ____________
Measure the object in 2 positions. ____________ ____________

Does the length change when the position changes?
Use pictures, numbers, or words to explain.

HOME CONNECTION

Your child can measure objects and distances in your home using their hand span, arm span, giant steps, or heel-to-toe steps. Encourage your child to estimate before measuring.

Focus | Children use a single copy of a unit to estimate and measure lengths in different positions.

Name: ______________________ Date: ______________________

Real-Life Beetles

The pictures match the size of each beetle in real life.
Measure the length of each beetle.

Choose a unit. ______________________
Use only 1 copy of the unit to measure.

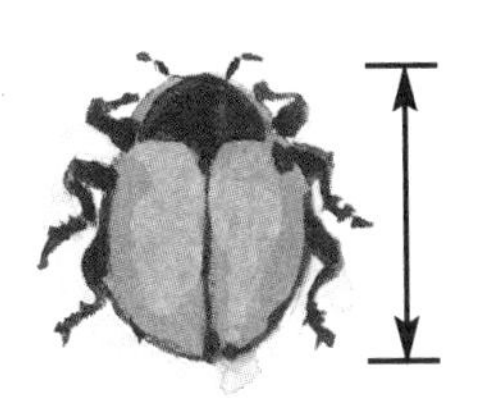

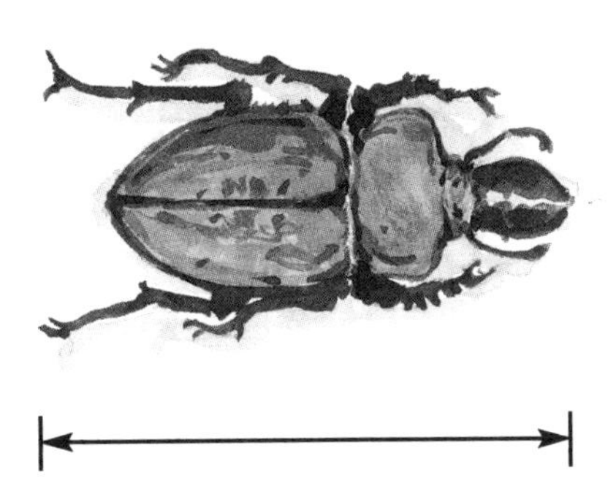

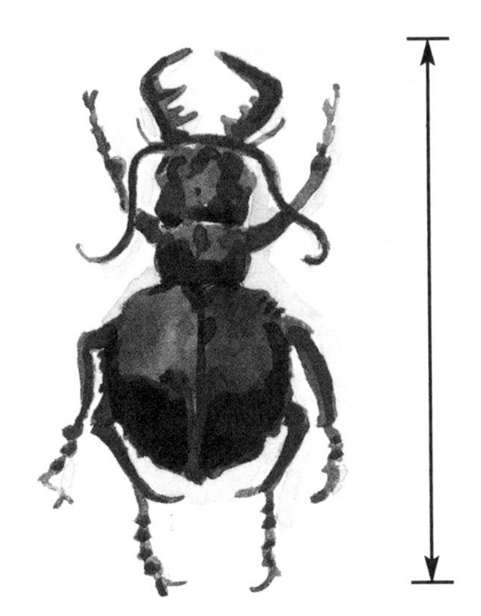

Which beetle is the longest?
Circle it.
Tell how you know. ______________________

____

FOCUS | Children measure, compare, and order the length of beetles using a single copy of a non-standard unit.

Name: ______________________ Date: ______________________

Distance Around

Choose 3 objects for measuring the distance around each.
Choose a unit. ____________________

What I measured	My estimate	My measurement
	about _______	
	about _______	
	about _______	

The distance around ____________ is shorter than the distance around ____________ and ____________.

____________ has the longest distance around.

Choose an object. ____________ Choose a unit. ____________

Measure the object in 2 positions. ____________ ____________
Does the distance around change when the position changes?
Tell how you know. Use pictures, numbers, or words.

HOME CONNECTION
Have children use non-standard units to measure around the wrists and ankles of different family members or friends, and then have them compare and order the measurements.

Focus | Children use a non-standard unit to measure distance around and compare and order measurements.

Name: ______________________ Date: ______________________

A Sum of 10

Choose a unit. ____________

Find a length, a height, and a distance around that total 10 units.

What I measured		My measurement
Length		
Height		
Distance around		

The sum of the measurements is ____ + ____ + ____ = ____ units.

Find a length between 20 and 30 units.

What I measured	The unit I chose	My measurement

Choose 1 of the problems.

Tell how you solved the problem. Use pictures, numbers, or words.

Focus | Children add measurements in the same non-standard units to solve a problem.

Name: ______________________ Date: ______________________

A Difference of 3

Choose a unit. ________________

Find a length and a distance around that have a difference of 3 units.

What I measured		My measurement
Length		
Distance around		

The difference between the 2 measurements is

______ – ______ = ______ units.

Find a length that is shorter than 1 unit.

What I measured	The unit I chose	My measurement

Choose 1 of the problems.

Tell how you solved the problem. Use pictures, numbers, or words.

Focus | Children subtract measurements in the same non-standard units to solve a problem.

Name: ____________________ Date: ____________________

Choosing a Unit

Choose an object. Measure the mass twice.
Use a different unit each time.

What I measured	The unit I chose	My measurement

Are the 2 measurements different? Why or why not? ____________

____

Choose another object. Measure the mass twice. ________ ________
Use a different unit each time.

What I measured	The unit I chose	My measurement

Compare the masses of the 2 units you used.
Did you need fewer units when you used a heavier unit? __________

Tell why or why not. ______________________________

Choose 1 of the objects you measured.

Which object is it? ____________________
What unit would you choose next time? Tell about your thinking.

____

Focus | Children measure the mass of objects using non-standard units and decide which unit was a better choice for measuring a given object.

Name: ______________________ Date: ______________________

About How Many?

Choose an object. ______________________
Estimate and measure the mass.
Use 3 different units.

The unit I chose	My estimate	My measurement
	about _______	
	about _______	
	about _______	

Tell how you chose the units. Use pictures, numbers, or words.

Choose 1 of the units you used. ______________
Measure the object in a new position. ______________
Does the mass change when the position changes? Tell why.

HOME CONNECTION 

Have your child estimate the masses of objects at home. For example, ask: "About how many of these spoons are as heavy as this bowl?"

Focus | Children estimate and measure mass using different non-standard units.

Name: ______________________ Date: ______________________

Balancing Act

Choose 3 objects for measuring masses
Choose a unit. ______________________
Estimate first.

What I measured	My estimate	My measurement
	about _______	
	about _______	
	about _______	

Show what you did in pictures, numbers, or words.

List the objects from lightest to heaviest.

______________ ______________ ______________

List the masses from heaviest to lightest.

______________ ______________ ______________

Focus | Children estimate, measure, and order the masses of objects.

Name: ______________________ Date: ______________________

Anna's Present Parcel

Measure the box for Anna's presents.

Choose a unit of measure. ____________________

Tell how you chose the unit. ____________________

Attribute	My estimate	My measurement
length	about _______	
height	about _______	

Tell how you estimated and measured.
Use pictures, numbers, or words.

Measure the distance around each present.

Anna's presents			
distance around			

Which present needs the longest ribbon? ____________________

Focus | Children choose a unit of measure and estimate, measure, compare, and order lengths.

Name: ______________________ Date: ______________________

Masses of Presents

Choose a unit of measure to measure the masses of the presents. ______________
Tell how you chose. ______________

What I measured	My estimate	My measurement
	about ______	
	about ______	
	about ______	

List the objects from heaviest to lightest.

______________ ______________ ______________

Do you need fewer units for a heavier unit? ______________

Tell why or why not. ______________________________

____

Can changing the position of a present make it lighter? __________

Tell why or why not. ______________________________

____

Write today's date. ______________

Anna's birthday is in 2 weeks.

What day and date is her birthday? ______________

Focus | Children choose a unit of measure and estimate, measure, compare the mass of objects. They relate the number of days to a week.

Name: ______________________ Date: ______________________

My Journal

Tell what you learned about measuring length, height, and distance around.
Use pictures, numbers, or words to show your thinking.

Tell what you learned about comparing how heavy objects are.
Use pictures, numbers, or words to show your thinking.

HOME CONNECTION 

Invite your child to practise measuring length, height, distance around, and mass while helping to purchase new clothes or recording their growth.

Focus | Children reflect on and record what they learned about length, height, distance around, and mass.

Addition and Subtraction to 100

Focus | Children identify the different ways to show 67.

Name: ______________________ Date: ______________________

Dear Family,

In this unit, your child will be developing strategies for adding and subtracting 2-digit numbers.

The Learning Goals for this unit are to

- Explore and develop personal strategies for adding and subtracting 1-digit and 2-digit numbers.
- Develop strategies to use when solving addition and subtraction problems up to 2-digits.
- Pose and solve a variety of number problems requiring addition or subtraction.
- Understand and use the relationship between addition and subtraction to help solve subtraction problems.
- Write a number as a sum or difference of 2 numbers.

You can help your child achieve these goals by doing the Home Connection activities suggested at the bottom of selected pages.

Name: ______________________ Date: ______________________

Take a Number

Choose a number from the team line-up. Colour the shirt red.

Show your number at least 2 ways.

Choose another number. Colour the shirt blue.
Show your number at least 2 ways.

How did you write the numbers in different ways?

____

____

Focus | Children represent numbers in different ways.

HOME CONNECTION 

If your child has a collection of objects at home – stickers, stuffed toys, rocks, models – gather some together and ask your child to count the collection in at least 2 different ways.

Name: ______________________ Date: ______________________

Dime Addition

Add dimes to each bank.
Write the total amount of money in each bank.

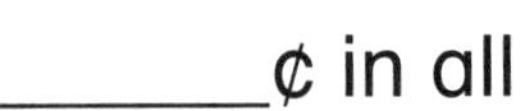
_______¢ in all

_______¢ in all

_______¢ in all

_______¢ in all

_______¢ in all

_______¢ in all

HOME CONNECTION 

Place 6 pennies in a row. Ask your child: "How many pennies are there?" Add a dime to the row and ask: "How much money is there altogether?" Continue adding dimes until you reach 96¢.

FOCUS | Children add dimes to find money amounts up to 99¢.

Name: ______________________ Date: ______________________

Adding 10s

Maria is playing a game about adding 10s.
She moves her counter on the 100-chart.
Use your 100-chart to see where she landed each time.

$26 + 10 =$ ____ $81 + 10 =$ ____ $18 + 70 =$ ____

$9 + 90 =$ ____ $31 + 40 =$ ____ $23 + 60 =$ ____

$$\begin{array}{r} 48 \\ +\ 30 \\ \hline \end{array} \quad \begin{array}{r} 17 \\ +\ 60 \\ \hline \end{array} \quad \begin{array}{r} 10 \\ +\ 40 \\ \hline \end{array} \quad \begin{array}{r} 15 \\ +\ 50 \\ \hline \end{array}$$

$29 + 30 =$ ____ $11 + 50 =$ ____ $16 + 80 =$ ____

$12 + 70 =$ ____ $36 + 40 =$ ____ $23 + 60 =$ ____

Suppose you add 20 to a number.
Predict how the number will change.

____

____

Suppose Maria started on 24 and landed on 74.
What did she add? ____

How could you check?

____

____

FOCUS | Children add groups of 10 to 1- and 2-digit numbers.

Name: ______________________ Date: ______________________

How Many Cakes?

The fair has an Ugly Cake contest for children.
There are 48 cakes entered in the contest.
Then, Mr. Melnik's class enters 6 more cakes.
How many cakes are there altogether? _____

How did you find the answer?
Use pictures, numbers, or words.

Now, find these sums.

8 + 6 = _____ 18 + 6 = _____ 28 + 6 = _____

38+ 6 = _____ 48 + 6 = _____ 58 + 6 = _____

Look at all the sums on this page. How are they the same?

____

____

Predict the answer to 68 + 6. Add to check.

Focus | Children use their own strategies to find the sums of 1-digit and 2-digit numbers. Then, they find related sums and look for patterns.

Name: ______________________ Date: ______________________

Addition Patterns

Find the sums.

9	19	29	39	49	59
+ 5	+ 5	+ 5	+ 5	+ 5	+ 5
___	___	___	___	___	___

What do you notice? ______________________

What is 79 + 5? ________
How do you know?

____

Find the sums.

4 + 17 = _____ 4 + 27 = _____ 4 + 37 = _____

4 + 47 = _____ 4 + 57 = _____ 4 + 67 = _____

What do you notice? ______________________

Find the sums.

13 + 0 = _____ 23 + 0 = _____ 33 + 0 = _____

43 + 0 = _____ 53 + 0 = _____ 63 + 0 = _____

What do you notice? ______________________

Focus | Children add 1-digit and 2-digit numbers and look for patterns in the sums.

Name: ______________________ Date: ______________________

Adding Animals

Choose an animal group. Make up an addition story.
Use materials to help. Show what you did.
Use pictures, numbers, or words.

Focus | Children add 2-digit numbers using their own strategies.

Name: ______________________ Date: ______________________

Find the Way Home

Help the cow find its way home to the barn.
Use any materials to find each sum.
Circle the pairs of numbers that add to 73.
Join the circled numbers to draw the path home.

64 + 9 = ____

56 + 9 = ____

57 + 17 = ____

50 + 23 = ____

27 + 18 = ____

40 + 23 = ____

37 + 25 = ____

42 + 31 = ____

32 + 41 = ____

43 + 31 = ____

62 + 10 = ____

28 + 45 = ____

Look at the pairs of numbers that add to 73.
Which sums were easy for you to find? Why?

____

____

Focus | Children add 2-digit numbers using their own strategies.

HOME CONNECTION 

Have your child explain how he or she found the answer to 64 + 9. Then, work together to think of 3 more pairs of numbers that add to 73.

Name: ______________________ Date: ______________________

Sports Day

On each turn, record the score for each member of your team.
Which unit did you use to measure? _____

Use base ten materials, 100-charts, or ten frames.
Find the total distance on each turn.

	Player A	Player B	Player C	Total Distance
Turn #1				_____ + _____ + _____ = ☐
Turn #2				_____ + _____ + _____ = ☐
Turn #3				_____ + _____ + _____ = ☐

Which turn was your team's best?
Show how you figured it out.
Use pictures, numbers, or words.

Focus | Children use personal strategies to add 3 numbers.

Name: ______________________ Date: ______________________

Yard Sale Day

You have 99¢.
What could you buy?
Work with a partner.
Choose 3 items.

Show how much you spent.	Choose 3 different items. Show how much you spent.
Draw the coins.	Draw the coins.
Write the number sentence.	Write the number sentence.
____ + ____ + ____ = ☐	____ + ____ + ____ = ☐

Could you buy 3 items with exactly 78¢? _____

Use pictures, numbers, or words to solve this problem.

HOME CONNECTION 

Label cans and packets of food with different prices from 1¢ to 33¢. Choose an amount between 50¢ and 99¢. Ask your child to find 3 items that together cost that amount.

Focus | Children create and solve addition problems.

Name: ______________________ Date: ______________________

Yard Sale Puzzles

Malik spent 88¢ at the yard sale.
What did he buy?
Write the number sentence. ______________________

Tell how you solved the problem.
Use pictures, numbers, or words.

What would you buy?

Write your yard sale puzzle here.

Tell some friends the total amount you spent.
Who can guess what you bought?

FOCUS | Children create then solve addition problems.

Name: ______________________ Date: ______________________

Hit the Target

Use a 100-chart. Start at 27.
Choose how many tens: 1 ten, 2 tens, or 3 tens
Flip a coin.
If it lands heads: Add that many 10s.
If it lands tails: Subtract that many 10s.
Record each turn.

27 ○ ______ = ______

______ ○ ______ = ______

______ ○ ______ = ______

______ ○ ______ = ______

______ ○ ______ = ______

______ ○ ______ = ______

How is subtracting 10s like adding 10s? How is it different?

Focus | Children add and subtract 10s, then compare their strategies for the 2 operations.

Name: ______________________ Date: ______________________

Take Tens

Teagan is playing a game.
On his 100-chart, he started at 83
and ended at 33.
Which number did Teagan subtract? _____

Use pictures, numbers, or words to solve this problem.

Teagan moved from 33 to 3.
What did he subtract this time? How do you know?

Use pictures, numbers, or words to show your work.

Add or subtract groups of 10 to make the sentences correct.
Use a 100-chart to help.

50 ◯ _____ ◯ _____ ◯ _____ = 50

35 ◯ _____ ◯ _____ ◯ _____ = 35

Focus | Children solve problems involving subtracting and adding 10s.

Name: ______________________ Date: ______________________

How Many Now?

There are 56 children in the Fun Run.
9 children stop for a drink.
How many children are still running? _____

Tell how you found the answer.
Use pictures, numbers, or words.

Here are other number stories about children in a Fun Run.

16	26	36	46
− 9	− 9	− 9	− 9
___	___	___	___

Look at the differences. How are they the same?

____

____

67 children are running. Make up your own subtraction problem.

____

____

Focus | Children use their own strategies to subtract 1-digit numbers from 2-digit numbers.
Then, they find related differences and look for patterns.

Name: ______________________ Date: ______________________

Subtraction Patterns

Find the differences.

8 – 6 = _____ 18 – 6 = _____ 28 – 6 = _____

38 – 6 = _____ 48 – 6 = _____ 58 – 6 = _____

What did you notice? ______________________

What is 78 – 6? _______
How do you know?

____

Find the differences.

14	24	34	44	54
– 8	– 8	– 8	– 8	– 8
___	___	___	___	___

What did you notice? ______________________

Find the differences.

99 – 0 = _____ 89 – 0 = _____ 79 – 0 = _____

69 – 0 = _____ 59 – 0 = _____ 49 – 0 = _____

What did you notice? ______________________

Focus | Children subtract 1-digit numbers from 2-digit numbers and look for patterns in the differences.

Name: ______________________ Date: ______________________

How Many More?

There are more dogs than rabbits entered in the show. How many more?
Show how you solved the problem. Use pictures, numbers, or words.

Focus | Children look for information in a story and subtract 2-digit numbers using their own strategies.

HOME CONNECTION 

Work with your child to write a different subtraction-story problem using the information about the pet show.

Name: ______________________ Date: ______________________

To the Fair

There are 38 Grade 2 children going to a country fair.
Some Grade 1 children are also going.
Altogether, 62 children will go.
How many Grade 1 children are going?

Tell how you solved the problem.
Use pictures, numbers, or words.

Make up your own subtraction story.

FOCUS | Children solve subtraction stories using their own strategies.

Name: ______________________ Date: ______________________

A Number Code

Ryan and Ayesha made up a code
that uses numbers in place of letters.
Here are some of the letters and numbers they use.

C	E	G	H	I	N	O	P	R	S	W
3	5	7	8	9	14	15	16	18	19	20

Ryan wrote these numbers: 16 9 7

Which animal name do they spell? ______________

Ayesha wrote subtraction sentences as clues for her favourite farm animal. Solve each clue.

34 – 15 = _____ The letter is _____.

67 – 59 = _____ The letter is _____.

49 – 44 = _____ The letter is _____.

61 – 56 = _____ The letter is _____.

78 – 62 = _____ The letter is _____.

What is Ayesha's favourite farm animal? ____________________

Which animal name do these differences spell? ______________

22 – 19 = _____ 28 – 13 = _____ 55 – 35 = _____

Focus | Children subtract 2-digit numbers using their own strategies to find encoded animal names.

Name: ______________________________ Date: ______________________________

Pocket Change

Takumi has 23¢ left.
How much could he have started with?
What did he buy?

93¢ 46¢ 54¢ 66¢ 50¢ 27¢ 88¢ 75¢ 21¢ 20¢ 30¢ 89¢ 65¢ 67¢ 13¢ 12¢ 18¢ 39¢

Use pictures, numbers, or words to show your work.

Find some classmates who did it a different way.
Make up your own problem about the yard sale.
Solve the problem.

Focus | Children find 2 numbers with a given difference and make up their own problem.

Name: ______________________ Date: ______________________

Missing Numbers

Use each number below only once.
Complete the subtraction sentences.

17

73

26

98

10

Use materials to help.
Draw a picture to show your thinking.

_____ − 59 = 14	51 − _____ = 41
36 − _____ = 19	_____ − _____ = 72

Now use the numbers to make up your own subtraction problem.
Give it to a classmate to solve.

____

____

Focus | From a set of given numbers, children choose those that complete subtraction sentences.

Name: ______________________ Date: ______________________

Feed the Pigs

Fern has forgotten how
many ears of corn she put in
Bucket 1 and Bucket 2.
How can you help her figure it out?

Fern used 95 ears of corn
altogether.
Show how you solved the problem.
Use pictures, numbers, or words.

Focus | Children find 2 numbers with a given sum.

Name: ______________________________ Date: ______________________________

What's Missing?

Oh no! Some of the Sports Day scores are missing.

TIGERS

	Suri	Kamal	TOTAL
Turn #1	23	14	
Turn #2	30		43
Turn #3		25	44

PANTHERS

	Antoine	Marc	TOTAL
Turn #1	10	29	39
Turn #2	23		51
Turn #3	37	8	

Use materials to help.

How far did Kamal's ball go on his second turn?

How far did Suri toss the ball on her third turn?

Marc's scores add to 65.
How far did his second toss go?

Focus | Children use personal strategies to find missing numbers in subtraction sentences.

Name: ______________________ Date: ______________________

Mystery Scores

Read the clues.
Find the other scores.

Bill's score is 59.

Lisa's score is 21 less than Bill's.
What is Lisa's score?

How do you know?

The difference between Karlie's score and Bill's score is 13.
Karlie's score is greater than Bill's score. What is Karlie's score?

How do you know?

Akif's score is 11 less than the sum of
Lisa's score and Bill's score.
What is Akif's score?

How do you know?

Focus | Children solve problems that involve finding a missing number in a subtraction sentence.

Name: ______________________ Date: ______________________

Showing 58

What *addition* story could have a sum of 58?
Draw a picture and write the number sentence.

58 = _____ + _____

_____ + _____ = 58

What *subtraction* story could have a difference of 58?
Draw a picture and write the number sentence.

58 = _____ − _____

_____ − _____ = 58

Are there more ways to get an answer of 58? Explain.

____

____

FOCUS | Children create and solve story problems that have a sum or difference of 58.

Name: ______________________ Date: ______________________

What's in Pat's Pocket?

Pat has 4 coins in her pocket.
She does not have any dimes.
The value of the coins is 20¢.
What are the coins? __________

Show how you solved the problem.
Use pictures, numbers, or words.

Pat has 4 coins in her pocket.
3 coins are the same.
The value of the coins is 76¢.
What are the coins?

Show how you solved the problem.
Use pictures, numbers, or words.

Focus | Children solve problems about 4 coins.

Name: ______________________ Date: ______________________

Name the Coins

There are 83¢ in Joe's piggy bank.
He empties his bank and counts the coins.
What could the coins be?

Show how to solve the problem.
Use pictures, numbers, or words.

HOME CONNECTION 

Take turns with your child creating money riddles. Model a money amount without showing it, then give the total and the number of coins.

Focus | Children choose a strategy to solve a problem.

Name: ______________________ Date: ______________________

At the Book Fair

Thina unpacked the 55 nature books.
She unpacked the 28 puzzle books.
How many books did Thina
unpack altogether? _______

How did you find the answer?
Use pictures, numbers, or words.

Rae put price stickers on all 78 picture books.
Jenn put price stickers on all 28 puzzle books.
How many more books did Rae price than Jenn? ______

How did you find the answer?
Use pictures, numbers, or words.

Focus | Children solve story problems involving addition and subtraction using personal strategies. They record their solutions pictorially.

Name: ______________________ Date: ______________________

Books on Order

	Monday	Tuesday	Wednesday
number of books ordered	26	43	79

Make up an addition problem about the books ordered.

____

____

Solve your problem.

Make up a subtraction problem about the books ordered.

____

____

Solve your problem.

Focus | Children create and solve story problems related to addition and subtraction of 2-digit numbers. They record their solutions pictorially and symbolically.

Name: ______________________ Date: ______________________

My Journal

Tell what you learned about adding.
Use pictures, numbers, or words.

Tell what you learned about subtracting.
Use pictures, numbers, or words.

Focus | Children reflect on and record what they learned about adding and subtracting.

Planning "Spring Fling"

Cam ran through the door. "I'm home!" he called out.
The dog came to see what the noise was about.
"We're having a 'Spring Fling'—a school celebration.
I've brought you your own special 'Fling' invitation!"

The next day Grade 2 started planning and thinking.
"What snacks will we serve? What will people be drinking?
How will we decorate? Where will we sit?"
Grandma thought, "Maybe I'll help out a bit."

Miss Chu said, “We’ll greet people when they arrive.
If everyone comes, there will be 75!
We’ll hand out the programs we make at the door.
We’ll keep extras here in this box on the floor.”

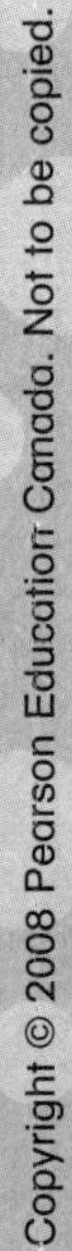

"We'll need a large space that can fit us all in.
We can hang up spring posters to show in the gym!
How will we set up the tables and chairs?
I hope there's enough. I hope we have spares!"

They brainstormed together for not very long.
"We should practise a dance, and maybe a song.
Reciting a spring poem would be special, too.
Are there any more things the Grade 2 class might do?"

They posted a chart that they studied a lot.
Cam asked, "Is there anything that we forgot?"
They grew more excited as every day passed.
Then finally the day came. They all sighed, "At last!"

They'd planned very well. They were all well prepared.
But they ran out of programs, so everyone shared.
The families grew quiet. The choir came in.
"Welcome," Cam said. "Let the 'Spring Fling' begin!"

About the Story

The story was read in class to prepare for a Mathematics Investigation activity. Children read a calendar, measured, and added and subtracted 2-digit numbers.

Talk about It Together

- Have you ever planned an event like "Spring Fling"? What are the kinds of things you had to think about? How did knowing math help you?
- Is there anything you would have done differently than the Grade 2 class?
- How were the children feeling as they planned the event?
- What is your favourite part of the story?

At the Library

Ask your local librarian about other good books to share about measuring and about using numbers.

Setting up the Classroom

Investigation 3

Suppose your class is setting up your classroom for 'Spring Fling.'

Choose a table or desk that you might want to move.

Choose a unit to measure its height and length.

____

Why did you choose that unit? ____________________

____

Measure the table or desk.

What is its length? ___________________________

What is its height? ___________________________

How do you know if the table or desk fits in the new place?
Use pictures, numbers, or words to show your thinking.

How Many Are Coming?

All the primary classes invited their families.
Some moms and dads are coming.
Some grandparents and friends are coming, too.

Count the tally marks for each class.
Write the number of guests in the last column.

Class	Tally of Guests	How Many Are Coming?
Kindergarten	卌 卌 卌 卌	_______ guests
Grade 1	卌 卌 卌	_______ guests
Grade 2	卌 卌 卌 卌 卌 卌 II	_______ guests

How many guests are coming altogether? _________
Show how you solved the problem.

Find someone who solved the problem in a different way.
Show how he or she solved it.
Use pictures, numbers, or words.

Write a subtraction problem about the guests.

Show how to solve your problem.
Use pictures, numbers, or words.

Ask someone to solve your problem.
How are the solutions the same? How are they different?
Use pictures, numbers, or words to explain.

When Is Spring Fling?

What month is the ‘Spring Fling’? ______________________

What day of the week is the ‘Spring Fling’? ______________

Cam’s class finished planning for the ‘Spring Fling’ on April 9.
Was this more than 1 week, 1 week, or less than 1 week before ‘Spring Fling’? ______________________________________

Tell how you know.

Cam’s class wants 2 to 4 days to decorate for ‘Spring Fling.’
What days might they start decorating? ________________
Tell how you know.

How Long Is It?

Find an object in your house that is about

- 3 footprints long
- 4 fingers high
- 2 arms long
- 5 hands high

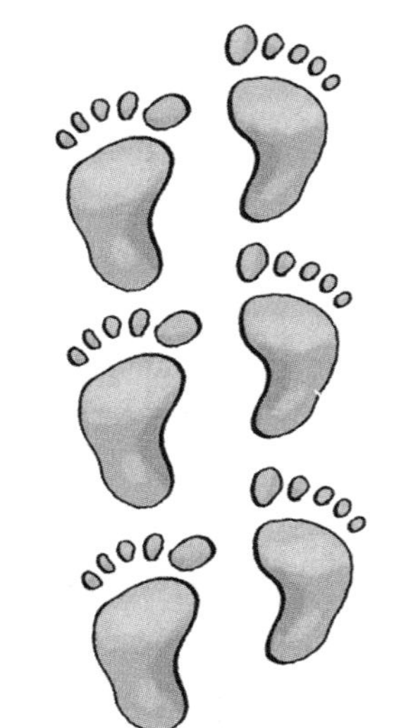

Guess first, then measure each object.

What do you think would happen if a grown-up looked for something 4 footprints long?

Creative Creatures

Suppose these 2 creatures changed into 1 new creature.
What number would the rods and cubes show?
(None can be taken away or added.)

 The next 4 pages fold in half to make an 8-page booklet.

Fold

Moving Day

Ryan could move this box:

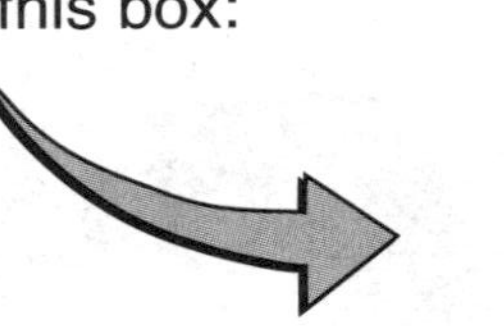

But he could not move this one:

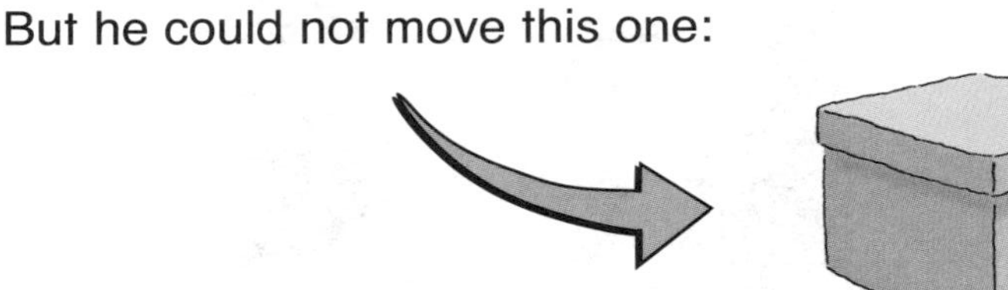

What do you think is in each box?

Your Favourite

Choose your favourite book.
Find another book that you think has a greater mass.
How could you check?
Find a third book that has less mass than your favourite book.
How do you know the mass is less?

What's the Difference? Game Board

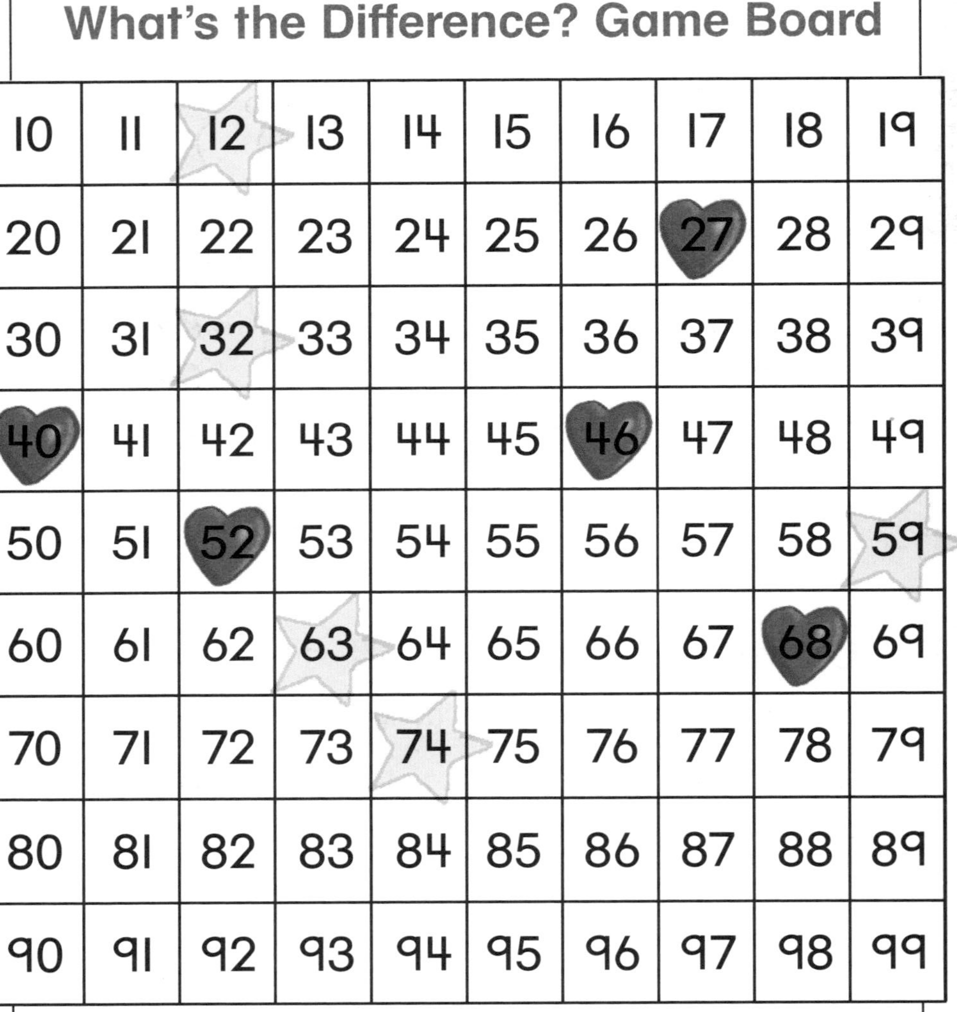

10	11	12	13	14	15	16	17	18	19
20	21	22	23	24	25	26	27	28	29
30	31	32	33	34	35	36	37	38	39
40	41	42	43	44	45	46	47	48	49
50	51	52	53	54	55	56	57	58	59
60	61	62	63	64	65	66	67	68	69
70	71	72	73	74	75	76	77	78	79
80	81	82	83	84	85	86	87	88	89
90	91	92	93	94	95	96	97	98	99

What's the Difference?

You'll need:
- game board (page 7)
- 2 sets of number cards, 1 to 9 and one 0 card, in a bag you cannot see through
- 6 counters
- paper
- pencil

On your turn:
- Draw 2 cards and use them to make a 2-digit number. Put a counter on the matching number on the game board.
- Draw 2 more cards and make another 2-digit number. Use a counter to cover this number on the game board.
- Find the difference between the 2 numbers by counting how many 10s and 1s they are away from each other. Use a counter to cover this number.

If the numbers **33** and **57** were covered, you would say, "57 is two 10s and four 1s away from 33. The difference is 24."

If the difference is

- more than 25, you get a point.
- a number with a ☆, you get a point.
- a number with a ♥, your friend gets a point.

Tally the points. Place the cards back in the bag. Remove your counters from the board.

Take turns until someone gets 10 points.

At the Fair

Game

What are some different ways you can win this carnival game?

Snack Time

You want to buy an apple.
You have this much money.

Do you have enough?

If not, how much more do you need?

Number Mystery

Use any digits you wish.
Find the greatest sum.

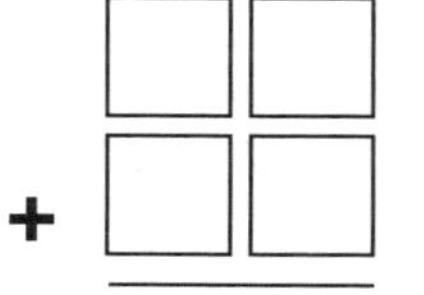

Suppose you wanted to make the least sum.
How would your thinking change?

Penny Measuring

Game

Find an object that you think would be
about 5 pennies long.
Use just I penny to check.
Now, place the object on this row of pennies.

Is the measurement the same? Why or why not?
Try it again with a different object.

Missing Numbers

Which digits are missing?
Is there more than one answer? Why?

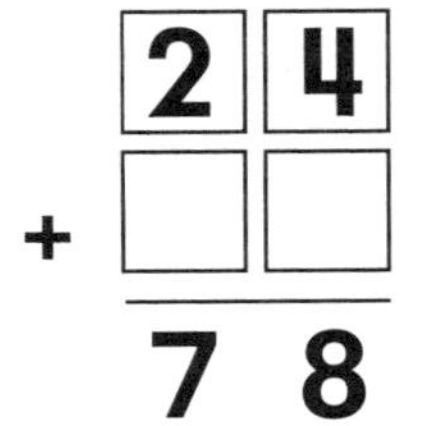

Make up your own for someone else to solve.
Who can you stump?

How High? How Far Around?

Gather I0 items of different shapes and sizes.
Order them from shortest to tallest.
Now, mix them up. Order them from greatest
distance around to least distance around.
Were the rows different? Why?

Calendar Romp

Suppose the date today is July 26.
Will we still be in the month of July in I week?

In 3 months, what will the month be?
6 months? One year?

UNIT 6

Geometry

Focus | Children identify and describe shapes and objects.

Name: ____________________ Date: ____________________

Dear Family,

In this unit, your child will be learning about 3-D objects, such as cubes, spheres, cylinders, cones, and pyramids, and 2-D shapes, such as triangles, squares, circles, and rectangles.

The Learning Goals for this unit are to

- Describe, compare, and sort 3-D objects according to their attributes, such as whether they roll or stack, have curved or flat parts.
- Describe, compare, and sort 2-D shapes according to their attributes, such as whether they have corners or straight sides.
- Make models to represent 3-D objects.
- Make models to represent 2-D shapes.
- Identify 2-D shapes on 3-D objects found in the environment.

You can help your child achieve these goals by doing the Home Connection activities suggested at the bottom of selected pages.

Name: ______________________ Date: ______________________

Our Construction

We made a ______________________ .

We used these objects.

Object	How Many Objects?
cube	
sphere	
cylinder	
cone	
pyramid	

We used _______ objects in all.

Focus | Children use 3-D objects in a construction and record the number of each type of object they used.

Name: ______________________ Date: ______________________

Look for It!

Colour the circles 

.

Colour the triangles 

.

Colour the squares 

.

Colour the rectangles Purple .

HOME CONNECTION 

With your child, look for examples in your home of some of the shapes your child has learned about, such as circles, triangles, rectangles, and squares. Which do you find most often?

Focus | Children look for and colour shapes in a drawing.

Name: ______________________ Date: ______________________

Same and Different

Circle 2 shapes.

I chose the ________________ and the ________________.

One way they are the same is ______________________

____.

Another way they are the same is ____________________

____.

One way they are different is ______________________

____.

Focus | Children select 2 shapes and describe how they are the same and different.

Name: ______________________ Date: ______________________

Build a Model

Use straws and clay to build a model.

I built a model for a ______________________.

My shape has ______________________.

Write 2 tips for building the model.

1. ______________________

____

____

____

2. ______________________

____

____

____

A ______________ cannot be built as a model with straws.

Why? ______________________

____

HOME CONNECTION

Ask your child: "What steps did you use to build the model?"

Focus | Children use straws and clay to build a model. They then give tips for building the model.

Name: ______________________ Date: ______________________

Use the Clues

Use geometric objects like these.

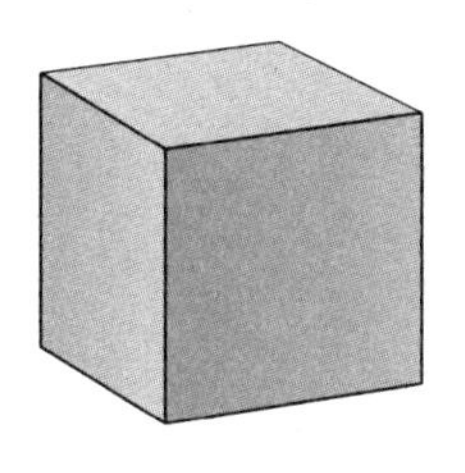

cube

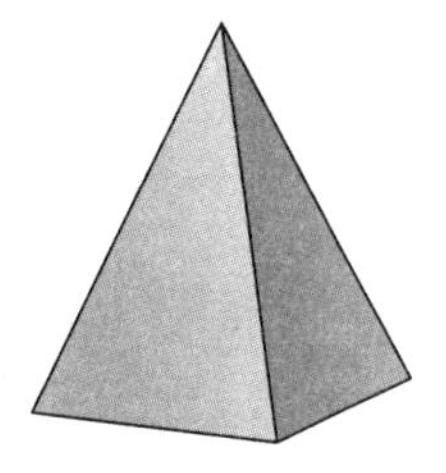

pyramid

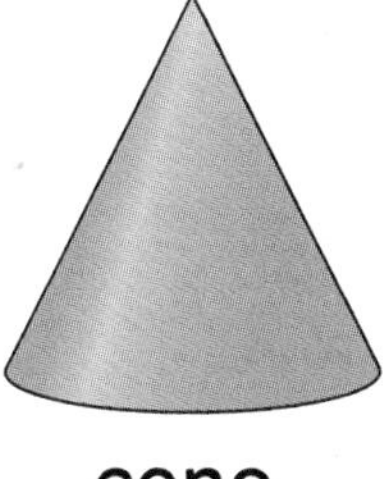

cone

cylinder

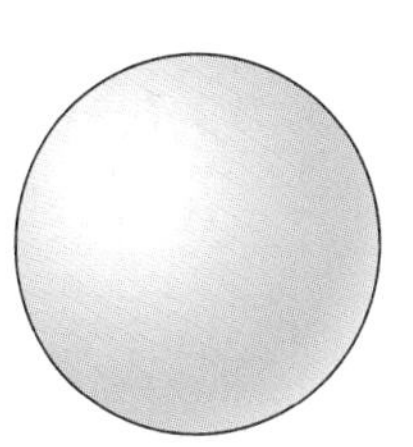

sphere

Follow the clues to fill in the chart.

Clue	Objects
all faces the same	
goes up to a point and it rolls	
flat faces and it rolls	
straight edges and some faces the same	
no vertices (corners)	

HOME CONNECTION 

With your child, hunt for objects that are examples of cubes, spheres, pyramids, cones, and cylinders. Have your child choose 2 objects and tell how they are the same and how they are different.

Focus | Children use clues to identify 3-D objects.

Name: ______________________ Date: ______________________

What Objects Do You See?

Write the name of the object you see in each drawing.

Focus | Children identify 3-D objects.

Name: ______________________ Date: ______________________

A Sorting Rule

Make a set of objects that have 2 things the same.

My sorting rule is ______________________________________

____.

Circle the objects that fit your rule.

What other way can you sort the objects?

Write another sorting rule. ______________________________

____

Put an ✗ on the objects that fit your new rule.

FOCUS | Children create 2 sorting rules and identify objects in the picture that fit each rule.

Name: ____________________ Date: ____________________

Objects That Are the Same

All these objects have ______________________.

They can all ______________.

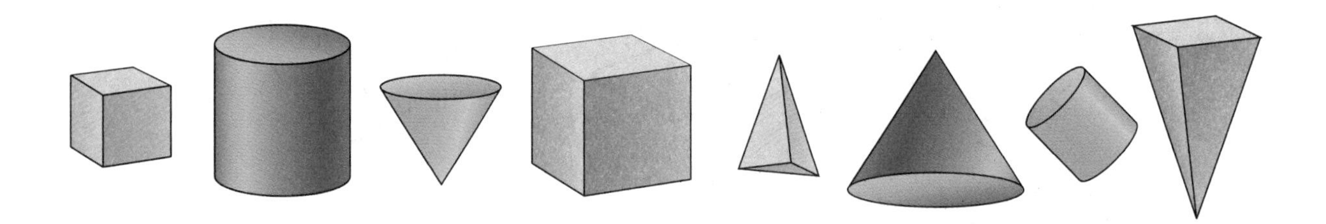

All these objects have ______________________.

They can all ______________.

Could you put a cube in the first set? Explain.

____

____

HOME CONNECTION 

Ask your child to think of an object that can be added to the first group and explain his or her choice. Repeat the activity for the second group.

Focus | Children identify common attributes of 3-D objects.

Name: ______________________ Date: ______________________

Making Objects

Use modelling clay to make one of these objects.

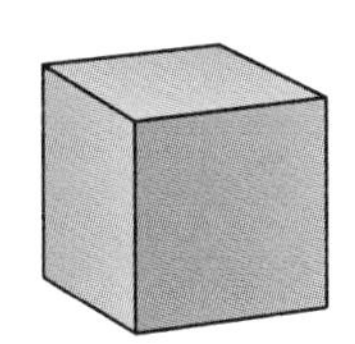 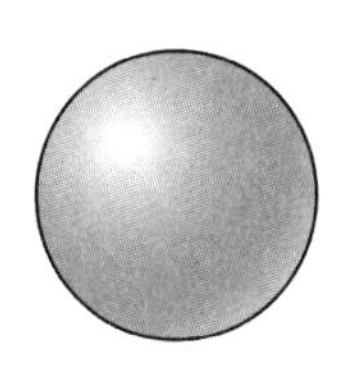 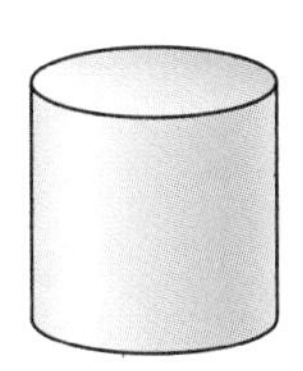  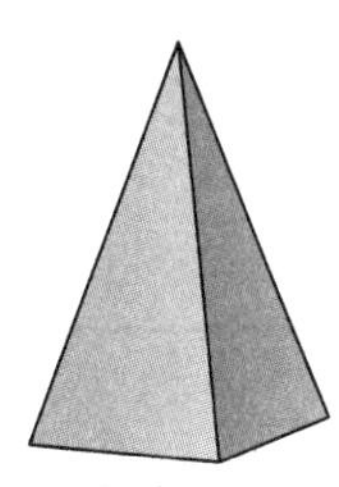

Name the object you made.
Show how you made your object.
Give tips that would help a friend make it.

Use pictures or words.

Which objects are best for building? Why? ______________________

____

Focus | Children construct an object from modelling clay and explain how they did it.

Name: ______________________ Date: ______________________

Where Am I?

Look in your classroom.
Find an example of each shape.
Tell where you saw each one.

square

rectangle

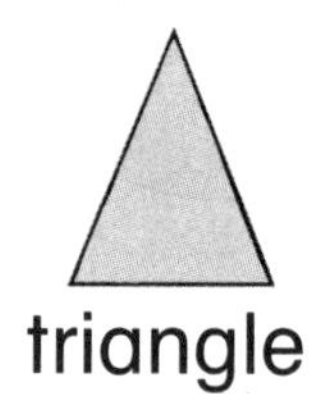
triangle

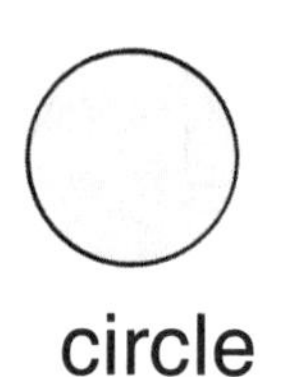
circle

What I Saw	It Looks Like	Where I Saw It
book cover	a rectangle	inside my desk

FOCUS | Children look for 2-D shapes in the classroom and tell where each is located.

Name: ______________________ Date: ______________________

What Shape Is the Face?

Choose an object in the classroom that fits on this page.
Draw or trace the circle, square, triangle, or rectangle face of the object.

What object did you choose? ______________________

It has a ______________________.

Focus | Children trace the face of an object found in the classroom and identify the shape.

Name: ______________________ Date: ______________________

What Is the Pattern Core?

Create the core of a repeating pattern.

The first object in the pattern core is round.

The second object in the pattern core has no corners.

The third object in the pattern core has straight edges.

All 3 objects are different.

What could the objects in the pattern core be?

Focus | Children choose a strategy to find 3 objects that make the core of a repeating pattern. There are multiple correct answers.

Name: ______________________ Date: ______________________

Making Patterns

Make a pattern. Use 6 ◯ and 3 ▭ .

Circle the pattern core.

Describe the pattern rule.

____

____

____

HOME CONNECTION 

Have your child use large red and small green circles to make a pattern in 2 different ways. Ask: "How are the patterns the same? How are they different?"

Focus | Children make a pattern using the instructions. There are multiple correct answers.

Name: ______________________________ Date: ______________________________

Pack the Spaceship

The astronauts need to pack these supplies in a bin but they can only take objects that have 2 attributes the same.

Choose a sorting rule for the bin.
Decide which objects to put in the bin.

Circle the objects that go in the bin.
These objects go in the bin because they all have ______________

____

Where do you see a triangle? ______________________________
Where do you see a square? ______________________________
Where do you see a circle? ______________________________
Where do you see a rectangle? ______________________________

Focus | Children sort objects according to 2 attributes and explain their reasoning.

Name: ______________________________ Date: ______________________________

Build a Spaceship

Which objects will you need to build this spaceship?

Make them from modelling clay.
Put them together to match the picture.
Which objects did you use to build the spaceship?

____

Are there any objects you did not use to build the spaceship? ____________________

What shapes do you see in the picture?

Shape	**Where did you see it?**
square	spaceship

Focus | Children record the objects they used to build a spaceship. Children record the shapes they see in the picture.

Name: ______________________ Date: ______________________

My Journal

What did you learn about objects?
Use pictures or words.

What did you learn about shapes?
Use pictures or words.

Focus | Children write what they learned about 3-D objects and 2-D shapes.

HOME CONNECTION 

Invite your child to help put away groceries. Talk about why you place particular packages or supplies together.

Data Analysis

Focus | Children talk about different ways to organize musical instruments.

Name: ______________________ Date: ______________________

Dear Family,

In this unit, your child will be learning about graphs, and about asking questions to gather information!

The Learning Goals for this unit are to

- Compare, sort, and organize real objects into concrete graphs.
- Use symbols to create pictographs.
- Read graphs and talk about the information on the graphs.
- Ask questions, collect data, and record the data.

You can help your child reach these goals by doing the suggested Home Connection activities at the bottom of certain pages.

Name: ______________________ Date: ______________________

What Different Ways Can We Sort?

Choose some pictures of musical instruments.
Sort the pictures into 2 sets in different ways.
What is the sorting rule of the last set you created?

Use the pictures to show your sorting.
Glue your pictures in rows or columns.

HOME CONNECTION 

Ask your child to sort some things into 2 sets. Help him or her line up the sets to match each object in 1 set with an object in the other set.

Focus | Children sort the same things in different ways.

Name: ______________________ Date: ______________________

What Does My Graph Tell Me?

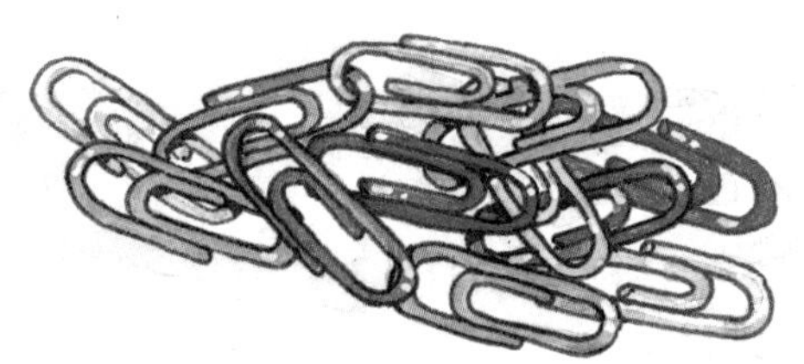

Number of Paper Clips

Silver paper clips										
Coloured paper clips										

What does your graph tell you?

____

____

____

Compare your graph to a friend's graph.

How are the graphs the same? ______________________

____

How are the graphs different? ______________________

____

Focus | Children create, interpret, and compare concrete graphs with 2 rows.

Name: ______________________ Date: ______________________

Which Colour Do You Like Best?

Choose 3 colours. Label the concrete graph.
Ask friends which colour they like best.
Create the graph.

Title: ______________________

________					
________					
________					

Which colour did most of your friends like? ______________

Which colour did fewest of your friends like? ______________

How is your graph the same as other graphs? ______________

____

Focus | Children create, interpret, and compare concrete graphs with more than 2 rows.

Name: ______________________ Date: ______________________

Dinosaur Graph

Ari's Toy Dinosaurs

orange

blue

red

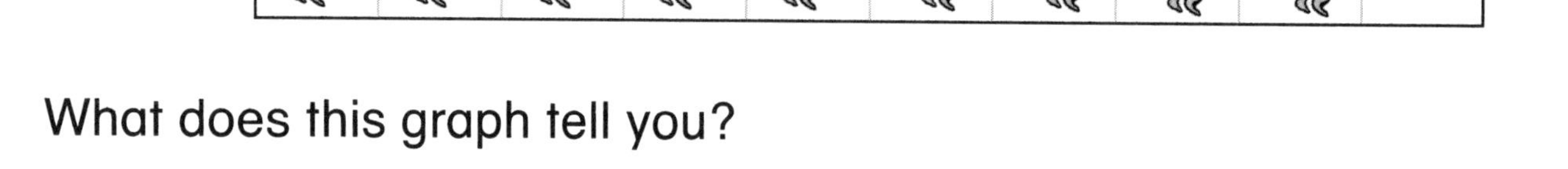

What does this graph tell you?

____

____

What 2 questions can you ask about the graph?

____

____

____

What are the answers to your questions?

____

____

____

Focus | Children read, interpret, and ask questions about a pictograph.

Name: ______________________ Date: ______________________

Which Do You Like Best?

Which sandwich do most people like best?

____

Which sandwich do fewest people like best?

____

____

How would the graph change if 3 more people said they like ham and cheese sandwiches best?

____

____

Focus | Children interpret a pictograph and describe how the graph changes.

HOME CONNECTION 

Have your child provide reasons for the change in the graph after his or her favourite sandwich is added.

Name: ______________________ Date: ______________________

Favourite Vegetables

Ask 5 friends to choose their favourite vegetable.
Cut and paste symbols to make a graph.

Which is Your Favourite Vegetable?

carrots					
celery					
tomato					

Which is the favourite vegetable? Tell how you know.
Use pictures, numbers, or words.

Focus Children make a pictograph and interpret the graph.

HOME CONNECTION 
Have your child ask family members to name their favourite vegetables, and then create a similar graph using matching hand-drawn pictures.

Name: ______________________ Date: ______________________

Deer on Cecile's Camping Trip

Cecile made a concrete graph to show the number of deer she saw on a camping trip.
She saw 4 deer on Monday and 5 deer on Tuesday.
She saw 13 deer altogether on Wednesday and Thursday.
She saw 5 more deer on Thursday than on Wednesday.

Make Cecile's concrete graph.

How many deer did Cecile see on Wednesday?

____

How many deer did Cecile see on Thursday?

____

Tell how you solved the problem.
Use pictures, numbers, or words.

Make a question about your graph.

____

____

Focus | Children solve a problem about a concrete graph, then record information about the graph.

Name: ______________________ Date: ______________________

Birds on Marco's Camping Trip

Marco used tallies to show the number of all the birds he saw.

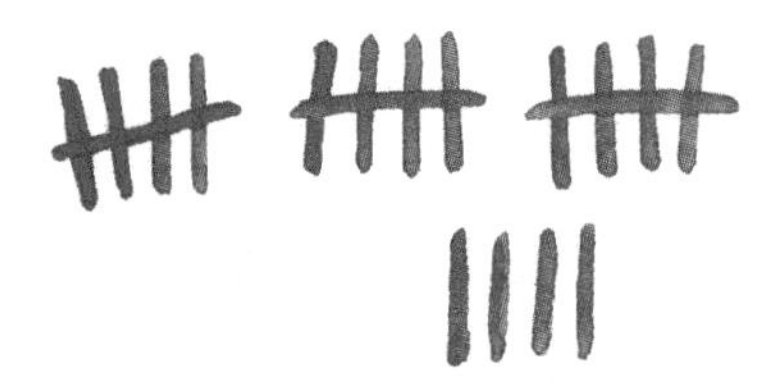

Marco used symbols to show the loons he saw.
He knows he saw 3 more hawks than eagles.
Finish Marco's graph.

Ask a question about your graph.

____

____

____

____

What is the answer to your question?

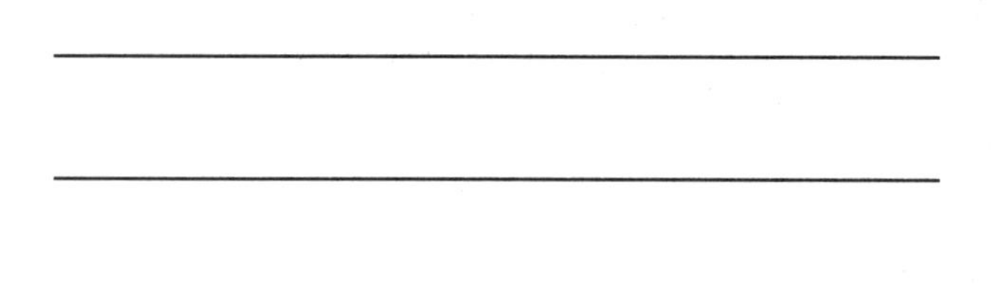

____

____

____

____

Birds on Marco's Camping Trip

loons eagles hawks

Focus | Children solve a problem about a pictograph, then ask a question about it.

Name: ______________________ Date: ______________________

Collecting Data

My Question ______________________

Ask 10 friends your question.
Use tallies or checkmarks to record your data.

What did you find out?

____

____

What do you think would happen if you asked the same question in a Grade 5 class? Why?

Focus | Children ask a question and organize the data using tallies or checkmarks.

HOME CONNECTION 

Have your child ask a question of the family, such as: "Do you like snow?" Have your child record the answers of up to 10 family members or friends using tallies or checkmarks.

Name: ______________________ Date: ______________________

What Is Your Favourite?

My Question

Do you like ____ ?

Ask 10 friends your question.
Record your data using tallies or checkmarks.

What did you find out? ______________________________________

____

What did you learn about collecting data?

____

____

Focus | Children ask a question and organize the data using tallies or checkmarks.

Name: ______________________ Date: ______________________

My Data

My Question ______________________________________?

Write a title. Label each column.

Ask friends your question.

Print their names in the columns.

Title: ______________________________________

_____________ _____________ _____________ _____________

What does your chart tell you?

____

____

HOME CONNECTION

Have your child ask up to 10 family members or friends a question, such as: "Do you like to skate, swim, or ride bikes?" Together, record the results in a chart.

Focus | Children collect data and record the results.

Name: ______________________ Date: ______________________

How Many Letters in Our Names?

Ask 6 friends to write their names where they belong.

fewer than 5 letters	5 letters	more than 5 letters

How many names have fewer than 5 letters? ________

How many names have 5 letters? ________

How many names have more than 5 letters? ________

Tell a partner a number story about your data.

HOME CONNECTION 

Have your child collect data about names using names of family members or other people they know.

Focus | Children record the number of letters in their friends' names and interpret the results.

Name: ______________________ Date: ______________________

Asking Questions

What problem will you solve by asking a question?

What question will you ask?

Record the data as you collect them.

What did you find out by asking your question?

Focus | Children create a question to gather information, pose the question, and record the results.

HOME CONNECTION

Help your child ask a question to gather information from friends and family members.

Name: ______________________ Date: ______________________

My Journal

What did I learn about collecting data and graphing?
Use pictures, numbers, or words.

Focus Children reflect on and record what they learned about collecting data and graphing.

HOME CONNECTION 

Have your child choose a favourite activity from the unit and tell you why it was a favourite.

The Field Trip

The Grade 2 class was so excited.
Even Grandma was invited.
Field Trip Day was finally here.
Everyone would visit Aquarium Place this year.

Inside they hardly made a sound.
There were glass and water tanks all around.
And so many different fish were there.
All colours and sizes—they were everywhere!

A whale swam by and looked their way.
"Welcome," its expression seemed to say.
Over in the corner they saw a crowd.
"A shark!" they suddenly gasped out loud.

Then some scuba divers came
and swam with fish that seemed quite tame.
"That looks like Grandma!" said Cam's friend Ben.
Cam blinked his eyes and looked again.

“I wonder,” said Cam as he stopped to stare.
“Just how many fish might fit in there?
How many tubs of water, too?”
“I’ll bet there are more than one hundred!” cried Lu.

Soon it was time for the dolphins to eat.
Buckets of fish were the lunchtime treat.
Some helpers held fish out over the tanks,
and dolphins jumped up and chirped out, "Thanks!"

What an amazing day it had been.
There were all kinds of undersea life they had seen.
They lined up again to get back on the bus, as they
wondered, "What did the fish think when they looked at *us*!"

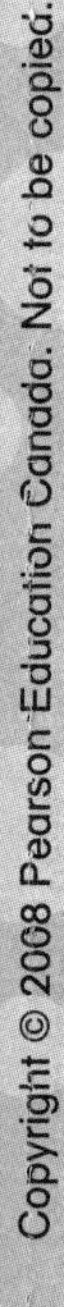

About the Story

The story was read in class to prepare for a Mathematics Investigation activity. Children made models of 3-D objects and 2-D shapes, created graphs, and answered questions about their graphs.

Talk about It Together

- What is your favourite part of the story?
- What is Grandma doing while the children look at all the fish in the tanks?
- Do you think working at Aquarium Place would be interesting? Why? Why not?
- What kind of jobs would the aquarium workers do?

At the Library

Ask your local librarian about other good books to share about 3-D objects and 2-D shapes, and about gathering and recording data.

Turtle Tank

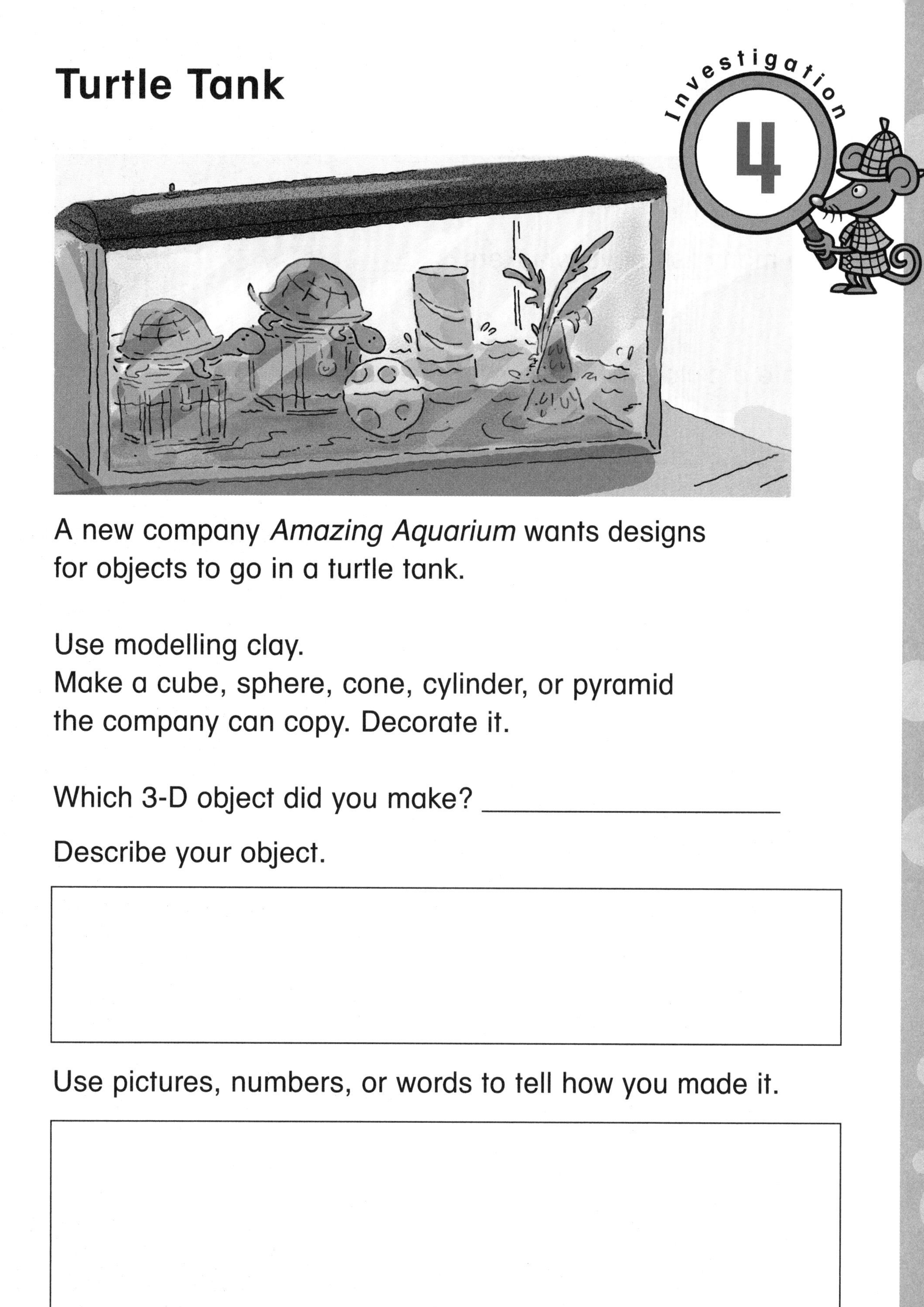

A new company *Amazing Aquarium* wants designs for objects to go in a turtle tank.

Use modelling clay.
Make a cube, sphere, cone, cylinder, or pyramid the company can copy. Decorate it.

Which 3-D object did you make? ____________________

Describe your object.

Use pictures, numbers, or words to tell how you made it.

A Fishy Graph!

How many fish have stripes? _____

How many fish have spots? _____

How many fish have whiskers? _____

Create a pictograph to show the results.

Title ______________________________

_____ _____ _____

What does your graph show about the fish?

Are there more striped or spotted fish in the tank? ________

How many more? ____________________________________

Show how you solved the problem.
Use pictures, numbers, or words.

Write your own problem about your graph.

Show how to solve your problem.
Use pictures, numbers, or words.

Making a Hoop for a Fish Tank

The company *Amazing Aquarium* wants hoop designs for a fish aquarium.

Use materials to make a triangle, square, rectangle, or circle hoop the company can copy.

Which 2-D shape did you use for your hoop? ____________
Describe your hoop.

Draw a picture of your hoop.

Egg Carton Graph

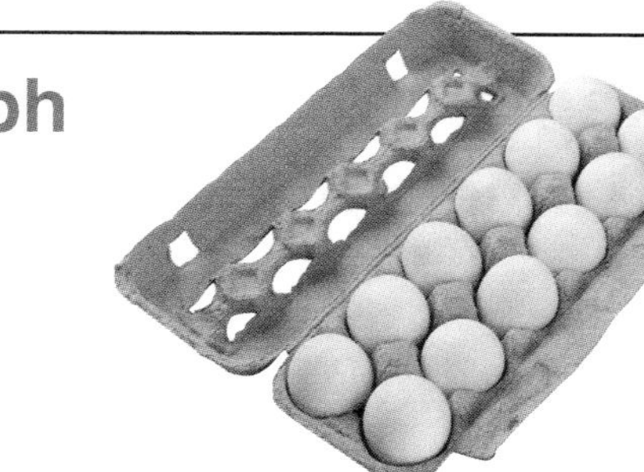

You'll need:
- a counter (coin, button)
- graph outline (below)
- an empty egg carton

Before you begin, colour each cup in the carton.
Your choices are red, green, blue, or yellow.

To play:
- Place the counter inside the egg carton and close it.
- Shake the carton, then open it.
- Record which colour egg cup the counter landed in using a symbol of your choice.
- Close the carton and shake it again.

After 10 turns, look at your graph.
Tell 3 things that you notice about it.

Red	Green	Yellow	Blue

 The next 4 pages fold in half to make an 8-page booklet.

Fold

Math at Home

I love math
and math loves me!
It's not too hard to see.
For everywhere I seem to go,
math always follows me!

My cookie has 12 chocolate chips.
Its shape is nice and round.
See? Math *is* all around!

Math at Home 3

Face Search

Look around the house for objects that have faces that are squares, triangles, circles, or rectangles. When you find one, tally it below. Which type of face do you think you will see most? Why?

Square	Circle	Rectangle	Triangle

Which face did you find the least?

Sandwich Shapes

Get creative with your lunch.
Cut your sandwich a different way each day.
You might try

The sky's the limit!

Tell Me About This Shape Game Board

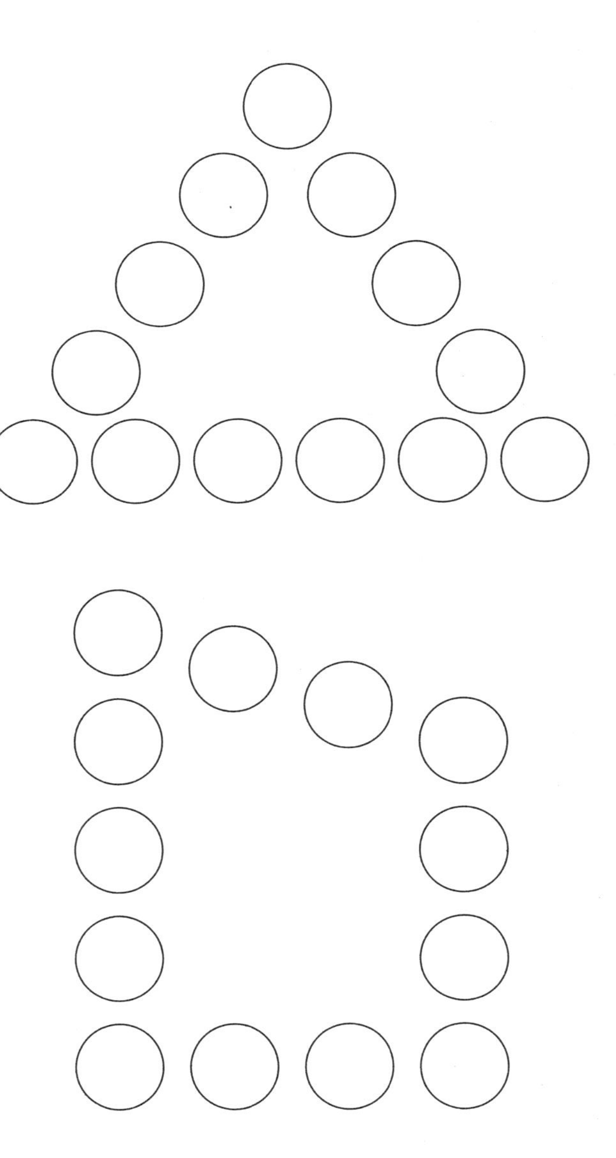

Tell Me About This Shape

You'll need:
- shapes in a bag you cannot see through

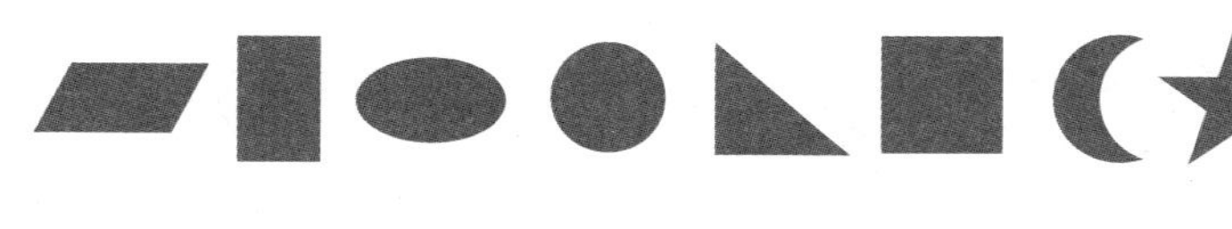

- small counters
- game board (page 7)

Before the game begins, each player chooses an outlined shape on the game board (page 7).

On your turn:
- Pull a shape from the bag. Place it between you and your partner.
- Both of you secretly write 1 thing that you notice about the shape. (Think about number of sides, lengths of the sides, number of corners.)
- Read your descriptions to each other.
  If they are **different**, you place 2 counters on your outlined shape.
  If they are the **same**, the other player places 1 counter on her outlined shape.

Take turns until someone's outlined shape is full.

Mystery Objects

Put 5 or 6 small 3-D objects into a bag you cannot see through.
Reach in and take 1 object.
Challenge a friend to guess your object by asking yes or no questions.
Only 10 questions are allowed.

If your friend guesses the object, show it.
Then give the bag to your friend and play again.
You guess this time!

Stick Shapes

Suppose it takes 3 craft sticks to make 1 side of a square.

How many craft sticks will it take to make the whole square?

Suppose it takes 5 craft sticks to make 1 side of another square.
How many craft sticks might it take to make the whole square?

Make other shape puzzles for a friend to solve.

Drop that Shape

Use a cardboard square.
Cut a piece of string the
same length as 1 of its sides.

Lay the square on a table in
front of a friend.
Tell your friend to pick up
the square and drop it on
the table.

When the square has been dropped,
ask your friend to use the string
to measure 1 side.
Did the length of the side change?

Now switch and let your friend hold
the square in the air.
Did the side length change?

What did you notice each time?

Do you think the same thing would happen
if you used a triangle?
How about a rectangle?
Find out.

Crazy Container Tally

Which types of food packaging
are most popular in your home?

Cans? Boxes? Jars?
Plastic containers?

Get ready to investigate
by printing each category
on a piece of paper.

The search is on! Tally
each one you find.

When you have finished, use all the information
to create a "Food Package" pictograph.

Tell someone about what you found.
Which was most popular? Least popular?
Were you surprised?

Out at the Park

Next time you are at the
park, look around at the
play equipment. What
shapes do you see?
Do you notice any
patterns?